Joseph D. Fehribach
Multivariable and Vector Calculus

Also of Interest

Complex Analysis
Theory and Applications
Bulboacă, Teodor / Joshi, Santosh B. / Goswami, Pranay, 2019
ISBN 978-3-11-065782-1, e-ISBN 978-3-11-065786-9,
e-ISBN (EPUB) 978-3-11-065803-3

Real Analysis
Measure and Integration
Markin, Marat V., 2019
ISBN 978-3-11-060097-1, e-ISBN 978-3-11-060099-5,
e-ISBN (EPUB) 978-3-11-059882-7

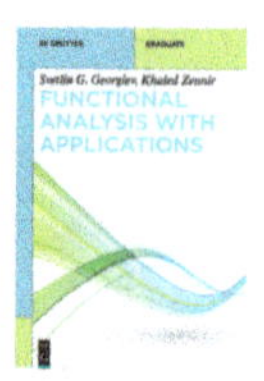

Functional Analysis with Applications
Georgiev, Svetlin G. / Zennir, Khaled, 2019
ISBN 978-3-11-065769-2, e-ISBN 978-3-11-065772-2,
e-ISBN (EPUB) 978-3-11-065804-0

Tensor Analysis
Schade, Heinz / Neemann, Klaus, 2018
ISBN 978-3-11-040425-8, e-ISBN 978-3-11-040426-5,
e-ISBN (EPUB) 978-3-11-040549-1

Applied Nonlinear Functional Analysis
An Introduction
Papageorgiou, Nikolaos S. / Winkert, Patrick, 2018
ISBN 978-3-11-051622-7, e-ISBN 978-3-11-053298-2,
e-ISBN (EPUB) 978-3-11-053183-1

Joseph D. Fehribach

Multivariable and Vector Calculus

DE GRUYTER

Author
Prof. Dr. Joseph D. Fehribach
Worcester Polytechnic Institute
Dept. of Mathematical Sciences
100 Institute Road
Worcester MA 01609-2280
United States of America
bach@math.wpi.edu

ISBN 978-3-11-066020-3
e-ISBN (PDF) 978-3-11-066060-9
e-ISBN (EPUB) 978-3-11-066057-9

Library of Congress Control Number: 2019951036

Bibliographic information published by the Deutsche Nationalbibliothek
The Deutsche Nationalbibliothek lists this publication in the Deutsche Nationalbibliografie; detailed bibliographic data are available on the Internet at http://dnb.dnb.de.

Cover image: Green's Windmill, Sneinton (Wikimedia https://commons.wikimedia.org/wiki/File:GreensMill4.JPG)
Typesetting: VTeX UAB, Lithuania
Printing and binding: CPI books GmbH, Leck

www.degruyter.com

To all those who have been my teachers and mentors in mathematics, physics, and chemistry over the years, in particular:

- Sr. Ancilla Lynch
- Professor Diane E. Vance
- Fr. John U. Caskey
- Mr. William Z. Young II
- Professor William C. Sagar
- Professor William H. Owens
- Professor Marshall Wilt
- Professor Neil A. Eklund
- Professor Robert Piziak
- Professor Dorothy Nelms
- Mr. Lewis D. Blake III
- Professor Gregory F. Lawler
- Professor Michael C. Reed
- Professor Barbara Lee Keyfitz
- Professor Michael Shearer
- Dr. Richard Ghez
- Professor David G. Schaeffer
- Professor John Chadam
- Professor Lambertus A. Peletier

Foreword

In the late 1990s, I was asked to teach our second-year course on vector and tensor calculus. When I looked at what books were available covering this material at roughly this level, I found that there were few choices. In fact, as best as I can now recall, I found only one: P. C. Matthews, *Vector Calculus*. Happily, this book covered exactly the topics in our course description, and at precisely the correct level. I adopted *Vector Calculus* for my class that year, and as far as I know, it has been used for the course ever since. With few exceptions, students from my classes that I have spoken to about the book have also liked it and found it helpful in learning the material.

About 10 years later, as I was teaching our course on multivariable calculus (which is typically the course that WPI students take before perhaps taking vector and tensor calculus), it occurred to me that there should be a book covering this multivariable material that was the counterpart to *Vector Calculus*. When I checked, I could not find such a book, so, eventually, I proposed to write it myself. The current text is the result. It is similar in spirit, though perhaps somewhat more rigorous mathematically (more emphasis on proofs) than the book that inspired it.

This book covers the material normally included in an American multivariable and vector calculus course. It is written, I hope, at a relatively high level, designed for students who have earned high marks on the AP Calculus BC exam (American system) or a maths A-level (British system) and who are interested in learning multivariable calculus in some depth. Some of the examples and exercises involve intricate (messy) calculations, and some are challenging (hard). Nonetheless, much of the material is straightforward, and there are many basic examples and relatively easy exercises. As is the case for Professor Matthews' text, this book is an alternative to the standard thousand-page calculus text that covers all of classical calculus with many applications and much discussion of ancillary material. Because this text is relatively brief, there is no remedial material, no discussion of numerical analysis, and there is no extensive treatment of applications (though several connections to physics are highlighted). It can be used as the text for a course or for self-study by students working on their own. But in any case, I hope that it is useful to students trying to master multivariable and vector calculus.

I wish to thank everyone who helped make this book possible, paricularly those who proof-read this work. No doubt some errors have made their way into print, though; anyone who finds errors is welcome to identify them to the author.

Joseph D. Fehribach
Worcester, Massachusetts
August 2019

https://doi.org/10.1515/9783110660609-201

Notes to the reader

Readers should be aware of the following points as they work through this book:

- Through out the book, several bits of mathematical shorthand are included, in part to help the reader understand what they mean and how they are used:
  - iff: if and only if
  - DNE: does not exit
  - $\perp$: perpendicular
  - $||$: parallel
  - $\simeq$: approximately equal to
  - $\Longrightarrow$, $\Rightarrow$: implies
  - $\Longleftrightarrow$, $\Leftrightarrow$: is equivalent to (logically the same as iff)
  - $\forall$: for all, for every
  - $\exists$: there exists; for some (when the symbol starts a prepositional phrase)
  - $\emptyset$: the empty set (the set containing no elements)
  - $\mathbb{R}$: the real numbers
  - $\mathbb{Q}$: the rational numbers ("Q" is for "Quotient")
  - $\mathbb{Z}$: the integers (from the German "Zahlen" for whole numbers)
  - $\mathbb{Z}^+$: the positive integers
- There are three symbols for equality used here:
  - The symbol ":=" is read as "defined equal to" and is used in definitions to indicate that the new entity on the left of the colon is defined to be the previously discussed entity on the right.
  - The symbol "$\equiv$" is read as "identically equal to" and means that the two equalities on either side are always equal; so, for example, $\sin^2\theta + \cos^2\theta \equiv 1$ no matter what angle θ is chosen.
  - The standard equal sign "=" is used in all other cases; so $x = 2$ in one example, but x may take on other values in other discussions.
- If f is a real-valued function defined on a domain D that is a subset of $\mathbb{R}^n$, this arrangement is expressed in symbols as $f : D \subset \mathbb{R}^n \to \mathbb{R}$. If f is continuous on its domain, this is symbolized as $f \in C(D)$. If f is n times differentiable and if its nth derivative $f^{(n)}$ is itself continuous on its domain, then one writes $f \in C^n(D)$.
- If $Q \subset \mathbb{R}^n$ is a region (or indeed a set), then
  - ∂Q: the boundary of Q
  - Q°: the interior of Q ($Q^\circ = Q - \partial Q$)
  - $\overline{Q}$: the closure of Q ($\overline{Q} = Q \cup \partial Q$)

https://doi.org/10.1515/9783110660609-202

Contents

1 Introduction/Background

1.1 What is multivariable and vector calculus?

As its title indicates, this book discusses multivariable and vector calculus. Most or all students who use this book will be familiar with single-variable calculus where, typically, $y = f(x)$. Most books or courses on single-variable calculus cover topics such as single-variable limits, differentiation, integration, and frequently, sequences and series. All of these will be taken as basic and background to this book.

Concepts from single-variable calculus are widely used today. Many problems in science, engineering, social science, and elsewhere, however, do not fit within the boundaries of single-variable calculus. They involve, for example, temperature, that can depend on the location in three-dimensional space as well as time, or sales figures that can vary with products, store locations, population density, and time. In the case of temperature, a function T might depend on four independent variables: x, y, z, and t. Writing it another way, $T : \mathbb{R}^4 \to \mathbb{R}$, which is read "$T$ maps $\mathbb{R}^4$ into the real numbers." Often one does not want to concentrate on the view of T as a function, but rather to think of it as a dependent variable, that is, $T = T(x, y, z, t)$. So in this case, T is both a multivariable function and the dependent variable—this is one type of function that we will concentrate on in this book.

In another case, one might want to study the track of a particle traveling through space. Now the only independent variable might be time t, but the dependent variables are the particle's location in three-dimensional space. Then among several alternatives, one might choose to express this location as a vector function: $\boldsymbol{x} : \mathbb{R} \to \mathbb{R}^3$ or $\boldsymbol{x} = \boldsymbol{x}(t)$. Here, one can think of $\boldsymbol{x}$ as (1) the point where the particle is located in $\mathbb{R}^3$, (2) the location vector in three-dimensional space from the origin to the point where the particle is located, and (3) the vector function giving that location as a function of time. All of these views of $\boldsymbol{x}$ will be discussed in this text.

An overview of the structure or layout of topics in calculus is displayed graphically in Figure 1.1. In the upper left-hand corner of the diagram is single-variable calculus where $y = f(x)$. As mentioned above, all of the topics, concepts, and material in this area is taken as given for our present discussion. We focus on extending calculus to the functions inside the red polygon: multivariable functions, vector functions, and vector fields. Multivariable functions have several independent variables, but a single dependent variable, and are typically of the forms $z = f(x, y)$ or $w = F(x, y, z)$. Vector functions, on the other hand, have a single independent variable and a vector with several components as the dependent variables. Vector functions are typically of the form $\boldsymbol{x} = \boldsymbol{x}(t)$ and the independent variable is often time. Vector fields have both multiple independent variables and vectors as outcomes. The arrows in Figure 1.1 indicate roughly the flow of the discussion of the material: We will first extend the concepts of calculus to vector functions and multivariable functions, then combine all of this to understand calculus and vector fields.

https://doi.org/10.1515/9783110660609-001

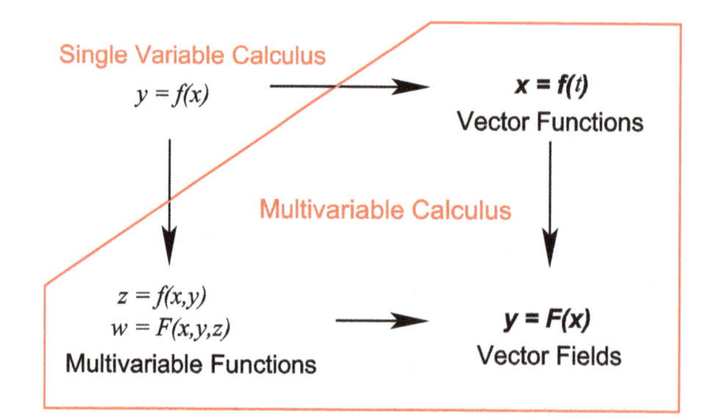

Figure 1.1: General overview of calculus. As one moves across this diagram, the number of domain variables increases; as one moves downward, the number of range variables increases. The diagram shows the main areas of calculus; those discussed in this book are inside the red polygon.

1.2 Vectors, lines, and planes in $\mathbb{R}^3$

This short section deals with several basic concepts in three-dimensional space ($\mathbb{R}^3$). While nothing in this section involves calculus, a clear understanding of this material is necessary to study multivariable calculus. Students familiar with the basics of these concepts, particularly in regard to vectors, may wish to skip this section or at least the first subsection.

1.2.1 Vectors

What is a vector? To a mathematician, the answer is that a vector is an element of a vector space, but this of course leads to an obvious question: "What is a vector space?" The answer to this second question is in any good linear algebra text, but for our proposes, the traditional physics answer to the question "What is a vector?" is probably most helpful: A *vector* is a physical or mathematical entity (thing) that has both length and direction. In $\mathbb{R}^3$, vectors can be thought of as arrows starting at the origin (the point $(0, 0, 0)$ where the axes meet) and ending at specific points. So if the vector $\langle 4, -1, \sqrt{3}\rangle$ begins at the origin, then it ends at the point $(4, -1, \sqrt{3})$, while if the vector $\boldsymbol{a} := \langle a_1, a_2, a_3\rangle$ begins at the origin, then it ends at the point (a_1, a_2, a_3) (see Figure 1.2). The beginning point for a vector is called its *tail* and the ending point is called its *head*. Vectors are unchanged by a rigid translation; thus the vector $\boldsymbol{a}$ could also begin at the point $(1, 1, 1)$ and end at $(a_1 + 1, a_2 + 1, a_3 + 1)$. In other words, vectors are determined by their lengths and the directions they point, but not where they are placed in space. In general, the vector starting at (x_o, y_o, z_o) and ending at (x, y, z) is $\langle x - x_o, y - y_o, z - z_o\rangle$.

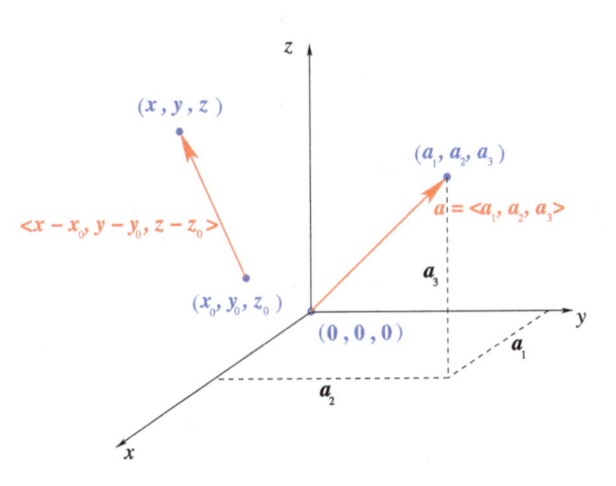

Figure 1.2: Coordinate axes, vectors (red) between points (blue) in $\mathbb{R}^3$. The vector from the origin to a point (a_1, a_2, a_3) is $\boldsymbol{a} := \langle a_1, a_2, a_3\rangle$; the vector from (x_o, y_o, z_o) to (x, y, z) is $\langle x - x_o, y - y_o, z - z_o\rangle$.

Another issue that will arise from time to time is how to name the coordinate axes. Following the standard tradition, the axes will normally be labeled as x and y in $\mathbb{R}^2$ and x, y and z in $\mathbb{R}^3$. But at times, it will be convenient to number the axes; thus in $\mathbb{R}^3$ the axes would be x_1, x_2 and x_3. This is particularly true if we are working in $\mathbb{R}^n$ for $n > 3$. The reader should become familiar with both of these conventions since both are used in various areas of mathematics, science, and engineering.

Throughout this section, let $\boldsymbol{a} := \langle a_1, a_2, a_3\rangle$ and $\boldsymbol{b} := \langle b_1, b_2, b_3\rangle$ be arbitrary vectors in $\mathbb{R}^3$. There are (at least) five basic definitions involving vectors that everyone studying vector calculus should know off the top of their heads:

Definition. The *length* of a vector $\boldsymbol{a} \in \mathbb{R}^3$ is defined as $|\boldsymbol{a}| := \sqrt{a_1^2 + a_2^2 + a_3^2}$.

Definition. *Vector addition*: The *sum* of $\boldsymbol{a}$ and $\boldsymbol{b}$ is defined as $\boldsymbol{a} + \boldsymbol{b} := \langle a_1 + b_1, a_2 + b_2, a_3 + b_3\rangle$. *Vector subtraction*: The *difference* between $\boldsymbol{a}$ and $\boldsymbol{b}$ is $\boldsymbol{a} - \boldsymbol{b} := \langle a_1 - b_1, a_2 - b_2, a_3 - b_3\rangle$. So this difference is the sum of $\boldsymbol{a}$ and the negative of $\boldsymbol{b}$.

Definition. *Scalar multiplication*: If c is a real number ($c \in \mathbb{R}$), then $c\boldsymbol{a} := \langle ca_1, ca_2, ca_3\rangle$.

Definition. The *scalar product*, *inner product*, or *dot product* (three names for the same product) of two vectors $\boldsymbol{a}, \boldsymbol{b} \in \mathbb{R}^3$ is defined as $\boldsymbol{a} \cdot \boldsymbol{b} := a_1b_1 + a_2b_2 + a_3b_3$.

Definition. The *cross product* of two vectors $\boldsymbol{a}, \boldsymbol{b} \in \mathbb{R}^3$ is given by expanding the following pseudo-determinant by its first row:

$$\boldsymbol{a} \times \boldsymbol{b} := \langle a_2b_3 - a_3b_2, a_3b_1 - a_1b_3, a_1b_2 - a_2b_1\rangle \equiv \begin{vmatrix} \mathbf{i} & \mathbf{j} & \mathbf{k} \\ a_1 & a_2 & a_3 \\ b_1 & b_2 & b_3 \end{vmatrix}$$

Remarks.

1. Of these five definitions, the first three are probably fairly straightforward. A vector's length is simply how long the arrow is. This definition of length is of course based on the Pythagorean theorem.[1] It is probably not so obvious why the definitions of the two products (inner and cross) are the correct ones, that is, the ones that match what nature and reality require. There are in fact other ways to define vector products, but as we shall see, these two arise naturally in many applications in math, physics, and engineering.
2. There is a secret word for how the basic vector operations other than the cross product work: *componentwise*. Vector addition, subtraction, scale multiplication, and the dot product all involve matching-up the various components of the vectors involved, or in the scalar multiplication case, doing the same thing to each component.
3. Vector addition has an important graphical implication: from the definition of addition, if the tail of vector $\boldsymbol{b}$ is placed at the head of vector $\boldsymbol{a}$, then the sum $\boldsymbol{a}+\boldsymbol{b}$ is the vector from the tail of $\boldsymbol{a}$ to the head of vector $\boldsymbol{b}$ as is shown in Figure 1.3.

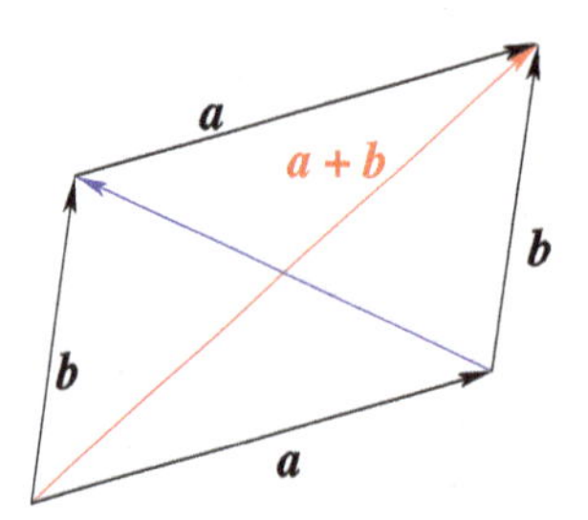

Figure 1.3: The sum of $\boldsymbol{a}$ and $\boldsymbol{b}$ is the red vector $\boldsymbol{a}+\boldsymbol{b}$ that runs from the tail of $\boldsymbol{a}$ to the head of $\boldsymbol{b}$. The diagram also makes clear the order that vectors are added does not change the sum: $\boldsymbol{a}+\boldsymbol{b}=\boldsymbol{b}+\boldsymbol{a}$. What is the blue vector in terms of $\boldsymbol{a}$ and $\boldsymbol{b}$?.

4. In the definition of the cross product, the array to the right of the identity sign, $\equiv$, is a pseudo-determinant. It is a pseudo-determinant (rather than a determinant) because its first row is made up of unit vectors rather than real numbers, and this is why it must be expanded by this row. The coordinate unit vectors are $\mathbf{i} := \langle 1,0,0\rangle$, $\mathbf{j} := \langle 0,1,0\rangle$ and $\mathbf{k} := \langle 0,0,1\rangle$ (see Figure 1.4). Anyone who is not familiar with determinants can either ignore this array and simply use the vector expression between the definition sign, :=, and the identity sign as the definition, or can look in almost any linear algebra text to see how a determinant is expanded.

1 Named for the Greek philosopher Pythagoras of Samos from the 6th century BC.

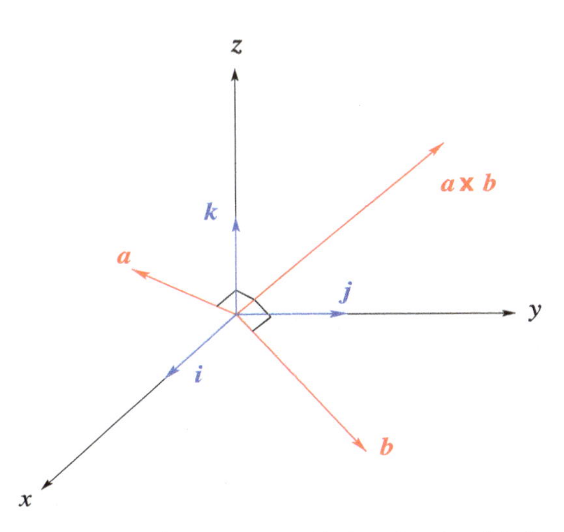

Figure 1.4: Coordinate axes, coordinate unit vectors (blue), and the cross product (red) in $\mathbb{R}^3$. As depicted here, the standard convention is that **i** points in the x-direction, **j** points in the y-direction, and **k** points in the z-direction. The vectors $\boldsymbol{a}$ and $\boldsymbol{b}$ are more-or-less randomly placed in this diagram, but given $\boldsymbol{a}$ and $\boldsymbol{b}$, the vector $\boldsymbol{a} \times \boldsymbol{b}$ is completely determined. Notice that the cross product $\boldsymbol{a} \times \boldsymbol{b}$ is perpendicular to both $\boldsymbol{a}$ and $\boldsymbol{b}$.

5. All the definitions above and the results below are basically the same in $\mathbb{R}^n$ for all $n \in \mathbb{Z}^+$, *except* for the cross product which is fundamentally a three-dimensional concept.[2]
6. Notice that the cross product of two vectors is itself a vector, but the dot product of two vectors is a *scalar*.

Example 1.1. If $\mathbf{x} = \langle 3, -2, 7\rangle$ and $\mathbf{y} = \langle -1, 2, 5\rangle$, it is easy to compute $|\mathbf{x}|$, $|\mathbf{y}|$, $\mathbf{x} + \mathbf{y}$, $\mathbf{y} - \mathbf{x}$, $\mathbf{x} \cdot \mathbf{y}$, $\mathbf{x} \times \mathbf{y}$:

Answer.

$$|\mathbf{x}| = \sqrt{3^2 + (-2)^2 + 7^2} = \sqrt{62}$$
$$|\mathbf{y}| = \sqrt{(-1)^2 + 2^2 + 5^2} = \sqrt{30}$$
$$\mathbf{x} + \mathbf{y} = \langle 3 + (-1), -2 + 2, 7 + 5\rangle = \langle 2, 0, 12\rangle,$$
$$\mathbf{y} - \mathbf{x} = \langle -1 - 3, 2 - (-2), 5 - 7\rangle = \langle -4, 4, -2\rangle,$$
$$\mathbf{x} \cdot \mathbf{y} = 3(-1) + (-2)(2) + 7(5) = 28$$
$$\mathbf{x}\times\mathbf{y} = \begin{vmatrix} \mathbf{i} & \mathbf{j} & \mathbf{k} \\ 3 & -2 & 7 \\ -1 & 2 & 5 \end{vmatrix} = [-2(5)-7(2)]\mathbf{i} - [3(5)-7(-1)]\mathbf{j} + [3(2)-(-2)(-1)]\mathbf{k} = \langle -24, -22, 4\rangle$$

The next proposition gives some basic results that begin to show why length, dot product, and cross product are so important.

2 There is, in fact, a cross product in $\mathbb{R}^7$ associated with the Fano plane, but this latter version is much less frequently used.

Proposition 1. *For any two vectors $\boldsymbol{a}, \boldsymbol{b} \in \mathbb{R}^3$, with θ the (smaller) angle between them:*
1. $\boldsymbol{a} \cdot \boldsymbol{b} = \boldsymbol{b} \cdot \boldsymbol{a}$
2. $|\boldsymbol{a}|^2 = \boldsymbol{a} \cdot \boldsymbol{a}$
3. $\boldsymbol{a} \times \boldsymbol{b} = -\boldsymbol{b} \times \boldsymbol{a}$
4. $\boldsymbol{a} \cdot \boldsymbol{b} = |\boldsymbol{a}||\boldsymbol{b}| \cos\theta$
5. $|\boldsymbol{a} \times \boldsymbol{b}| = |\boldsymbol{a}||\boldsymbol{b}| \sin\theta$
6. $|\boldsymbol{a} \times \boldsymbol{b}| \leq |\boldsymbol{a}||\boldsymbol{b}|$; $|\boldsymbol{a} \cdot \boldsymbol{b}| \leq |\boldsymbol{a}||\boldsymbol{b}|$ *(Cauchy-Schwarz inequalities)*
7. *The direction of* $\boldsymbol{a} \times \mathbf{b}$ *is perpendicular to the plane containing* $\boldsymbol{a}$ *and* $\mathbf{b}$, *and obeys the right-hand rule. This relationship is shown in Figure* 1.4.
8. $|\boldsymbol{a} + \boldsymbol{b}| \leq |\boldsymbol{a}| + |\boldsymbol{b}|$ *(triangle inequality)*

Remark. The *right-hand rule* is a mnemonic for remembering the direction of the cross product relative to the two vectors, as well as the standard orientation of the coordinate axes. According to the rule, if one points the fingers of one's *right* hand in the direction of a vector $\boldsymbol{a} \in \mathbb{R}^3$, then curls these fingers toward the direction of $\boldsymbol{b} \in \mathbb{R}^3$, the right thumb points in the direction of the cross product $\boldsymbol{a} \times \boldsymbol{b}$. This rule also gives the orientation of a right-handed coordinate system in $\mathbb{R}^3$: if one points the fingers of one's right hand in the x or x_1 direction, then curls the fingers toward the direction of y or x_2, the thumb points in the direction of z or x_3. Notice that if one carries out the above procedure with one's left hand, the thumb points in the opposite direction.

Proof. For (1), (2), and (3), each proof is simply the application of the definition and is left as an exercise (Exercise 1.8). For (4), the key is the law of cosines applied to the diagram in Figure 1.5 below. According to the law of cosines, $|\boldsymbol{a} - \boldsymbol{b}|^2 = |\boldsymbol{a}|^2 + |\boldsymbol{b}|^2 - 2|\boldsymbol{a}||\boldsymbol{b}| \cos\theta$. On the other hand, $|\boldsymbol{a} - \boldsymbol{b}|^2 = (\boldsymbol{a} - \boldsymbol{b}) \cdot (\boldsymbol{a} - \boldsymbol{b}) = \boldsymbol{a} \cdot \boldsymbol{a} - 2\boldsymbol{a} \cdot \boldsymbol{b} + \boldsymbol{b} \cdot \boldsymbol{b} = |\boldsymbol{a}|^2 + |\boldsymbol{b}|^2 - 2\boldsymbol{a} \cdot \boldsymbol{b}$. Equating these two expressions and cancelling, one finds the desired formula. The proofs of (5), (7) and (8) are computational and are also left as exercises (Exercises 2.9, 2.8 and 2.10). Finally (6) are versions of the Cauchy-Schwarz inequality that follow from (4) and (5) since sine and cosine are bounded between −1 and 1. □

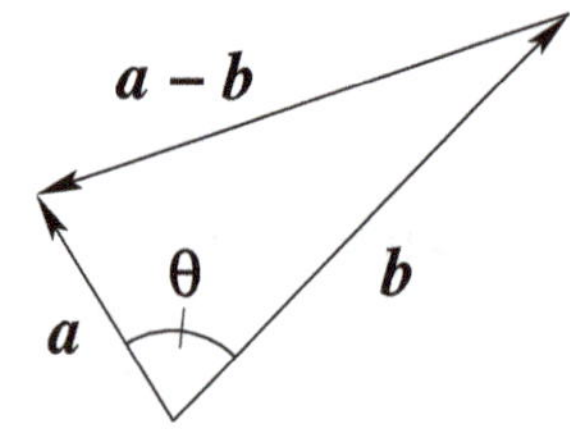

Figure 1.5: Diagram to determine the relationship between the dot product and the angle θ. By the law of cosines, the lengths of the three vectors must be related as $|\boldsymbol{a} - \boldsymbol{b}|^2 = |\boldsymbol{a}|^2 + |\boldsymbol{b}|^2 - 2|\boldsymbol{a}||\boldsymbol{b}| \cos\theta$.

Example 1.2. Suppose that $\boldsymbol{a} = \langle 1, -1, 3 \rangle$ and $\boldsymbol{b} = \langle 6, 2, -2 \rangle$. Please find the angle between these two vectors when they are placed tail-to-tail.

Answer. The angle between two given vectors can be found using either (3) or (5) from Proposition 1, but since dot products are easier to compute, (3) is the easier choice. Using (3), one finds that

$$\frac{\boldsymbol{a}\cdot\boldsymbol{b}}{|\boldsymbol{a}||\boldsymbol{b}|}=\frac{1(6)-1(2)+3(-2)}{\sqrt{1^2+(-1)^2+3^2}\sqrt{6^2+2^2+(-2)^2}}=\frac{-1}{11}=\cos\theta\,,$$

or that $\theta = \mathrm{Cos}^{-1}(-1/11) \simeq 1.662$ radians (about 95 degrees).

There are two important results that follow immediately from (3) and (5) in the previous proposition and help us determine whether vectors are perpendicular or parallel.[3]

Definition. Two nonzero vectors $\boldsymbol{a},\boldsymbol{b} \in \mathbb{R}^3$ are *perpendicular* ($\boldsymbol{a} \perp \boldsymbol{b}$) iff the angle between them is $\pm\pi/2$. These vectors are *parallel* iff the angle between them is zero. They are *antiparallel* iff the angle between them is $\pm\pi$. If $\boldsymbol{a}$ and $\boldsymbol{b}$ are either parallel or antiparallel, then $\boldsymbol{a}||\boldsymbol{b}$.

Corollary 1. *Suppose that* $\boldsymbol{a}$ *and* $\boldsymbol{b}$ *are both nonzero. Then*
- $\boldsymbol{a}\cdot\boldsymbol{b} = 0$ *iff* $\boldsymbol{a} \perp \boldsymbol{b}$ *(i. e.,* $\theta = \pm\pi/2$*).*
- $\boldsymbol{a}\times\boldsymbol{b} = 0$ *iff* $\boldsymbol{a}||\boldsymbol{b}$ *(i. e.,* $\theta = 0, \pm\pi$*).*

Example 1.3. For what value of x_1 are the vectors $\langle x_1, 6, -3\rangle$ and $\langle 5, -2, 7\rangle$ perpendicular?

Answer. This result follows from the first corollary above: vectors are perpendicular exactly when their dot product is zero. So the requirement is that

$$5x_1 + 6(-2) + (-3)(7) = 0 \quad\Longleftrightarrow\quad x_1 = 33/5\,,$$

which means that $\langle 33/5, 6, -3\rangle$ and $\langle 5, -2, 7\rangle$ are perpendicular.

1.2.2 Planes in $\mathbb{R}^3$

Next, we need to define something that most people have an intuitive sense of: a plane in $\mathbb{R}^3$.

Definition. Suppose that (x_o, y_o, z_o) is a fixed point in $\mathbb{R}^3$, and $\boldsymbol{N} := \langle A, B, C\rangle$ is a fixed vector. Then Π is the *plane* in $\mathbb{R}^3$ passing through (x_o, y_o, z_o) perpendicular to $\boldsymbol{N}$ iff when (x, y, z) is an arbitrary point on Π, the vector $\langle x-x_o, y-y_o, z-z_o\rangle$ is perpendicular

3 Is the zero vector perpendicular to other vectors? Perhaps surprisingly, the answer depends on which source one looks to. Matthews [3] only applies the terms “perpendicular” and “parallel” to nonzero vectors, and this is the view taken here. But some other authors (e. g., Kosmala [1]) consider the zero vector to be both perpendicular and parallel to all other vectors.

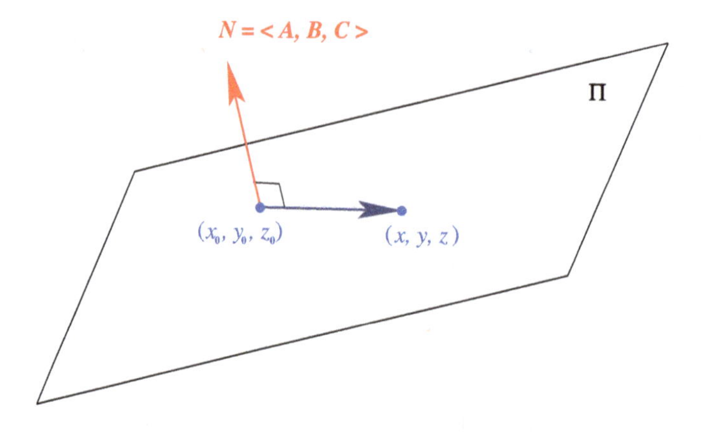

Figure 1.6: The point-normal equation of a plane. Here, $\boldsymbol{N}$ represents *any* vector normal or perpendicular to the plane, and $\langle x - x_o, y - y_o, z - z_o\rangle$ is a vector in the plane, starting at a fixed point (x_o, y_o, z_o) and ending at any point (x, y, z) on the plane. Since these vectors are always perpendicular, their dot product must be zero.

or normal to $\boldsymbol{N}$ (see Figure 1.6). In other words,

$$\boldsymbol{N} \cdot \langle x - x_o, y - y_o, z - z_o\rangle = A(x - x_o) + B(y - y_o) + C(z - z_o) = 0.$$

Remark. The formula in the definition ($A(x-x_o)+B(y-y_o)+C(z-z_o) = 0$) is sometimes called the point-normal equation of a plane. One can always collect the constant terms as $D = Ax_o + By_o + Cz_o$ to get the general equation of a plane: $Ax + By + Cz = D$.

Example 1.4. Find the equation of the plane perpendicular to $\langle 3, -4, 7\rangle$ that passes through the point $(1, 2, 6)$.

Answer. In this case, $\boldsymbol{N} = \langle 3, -4, 7\rangle$ and $(x_o, y_o, z_o) = (1, 2, 6)$. So the equation of the plane is simply $3(x - 1) - 4(y - 2) + 7(z - 6) = 0$ or $3x - 4y + 7z = 37$.

Example 1.5. Find the equation of the plane passing through the points $(1, 0, -1)$, $(0, -1, 1)$, and $(-1, 1, 0)$.

Answer. The key to solving this problem is to note that the three points can be used to form several different vectors that all lie in the plane. The cross product of any two of these vectors (that are not scalar multiples of each other) will produce a normal vector that can then be used with any of the three points in the point-normal equation of the plane.

So the vector $\boldsymbol{a}$ can be defined as $\langle 1 - 0, 0 - (-1), -1 - 1\rangle$, while $\boldsymbol{b}$ can be defined as $\langle 0 - (-1), -1 - 1, 1 - 0\rangle$. Then $\boldsymbol{N} = \boldsymbol{a} \times \boldsymbol{b} = \langle -3, -3, -3\rangle$. Taking $(x_o, y_o, z_o) = (1, 0, -1)$, one finds that

$$-3(x - 1) - 3(y - 0) - 3(z + 1) = 0 \quad \text{or} \quad x + y + z = 0.$$

Remark. An equivalent equation (or perhaps the same equation) would be obtained if $\boldsymbol{a}$ and $\boldsymbol{b}$ were defined using the given points in a different order, or if a different point in the plane were used as (x_0, y_0, z_0).

1.2.3 Lines in $\mathbb{R}^3$

Before studying even the most basic version of calculus, everyone is likely to be familiar with the equation that represents a line in the x, y-plane. Specifically any line in the x, y-plane that is *not* vertical can be represented by an equation of the form $y = mx + b$ where $m \in \mathbb{R}$ is its slope and $b \in \mathbb{R}$ is its y-intercept. For a horizontal line, the equation is $y = b$ which has this form with $m = 0$. For a vertical line, the equation is $x = a$ where $a \in \mathbb{R}$ is some fixed value; this equation is of course not in the slope-intercept form.

1.2.3.1 Vector form

Now we turn our attention to representing lines in $\mathbb{R}^3$. At first, one might expect again that this would be done by a single linear equation. A single linear equation, however, was shown in the previous section to represent a plane in $\mathbb{R}^3$, not a line. As we shall see, a line in $\mathbb{R}^3$ will require a vector equation, or equivalently, to three scalar equations.

Consider a line $\mathcal{L}$ in $\mathbb{R}^3$ depicted in Figure 1.7. Let $\boldsymbol{x}_o$ be any vector from the origin to $\mathcal{L}$, and let $\boldsymbol{m}$ be any vector of positive length lying along $\mathcal{L}$. Notice that $\boldsymbol{m}$ can be placed at the head of $\boldsymbol{x}_o$, as depicted in Figure 1.7, and that the line $\mathcal{L}$ is then traced

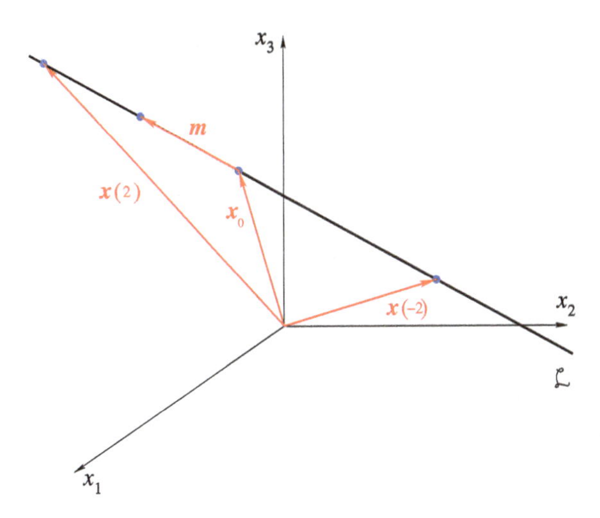

Figure 1.7: The use of vectors to represent a line $\mathcal{L}$ in $\mathbb{R}^3$. The vector $\boldsymbol{x}_o$ is from the origin to some fixed point on $\mathcal{L}$; the vector $\boldsymbol{m}$ lies along the line. The vector function $\boldsymbol{x}(t) = \boldsymbol{m}t + \boldsymbol{x}_o$ is from the origin to a point that sweeps out the line as t varies from $-\infty$ to ∞.

out by the vector equation

$$\boldsymbol{x}(t) = \boldsymbol{m}t + \boldsymbol{x}_o \tag{1.1}$$

as t runs through all real values.[4] In particular, $\boldsymbol{x}(0) = \boldsymbol{x}_o$, while for $t = 2$, the vector $\boldsymbol{x}$ has moved so that its head is now at the head of $2\boldsymbol{m}$ as in Figure 1.7. For $t = k$, where k is a positive integer, the vector $\boldsymbol{x}$ has moved so that its head is k lengths of $\boldsymbol{m}$ along the line $\mathcal{L}$. For $t = -k$, the vector $\boldsymbol{x}$ has moved k lengths of $\boldsymbol{m}$ in the opposite direction from the head of $\boldsymbol{x}_0$. The vector $\boldsymbol{m}$ is the direction vector for $\mathcal{L}$, and is the rough equivalent of the slope m in the slope-intercept form described above. This vector equation (1.1) is called the *vector form* of the equation of a line in $\mathbb{R}^3$.

Example 1.6. Find an equation in vector form for the line passing through the points $(1, 0, -2)$ and $(-1, 3, 1)$.

Answer. Among the many possible choices for how to define $\boldsymbol{x}_o$ and $\boldsymbol{m}$ in this case, let us pick $\boldsymbol{x}_o = \langle 1, 0, -2\rangle$ (based on the first of the two points), and $\boldsymbol{m} = \langle -1-1, 3-0, 1-(-2)\rangle$ (a vector connecting the two points). The equation of the line is then

$$\boldsymbol{x}(t) = \langle -2, 3, 3\rangle t + \langle 1, 0, -2\rangle .$$

Remark. Keep in mind that the equation in the previous example is *an* equation for the line; it is not *the* equation—it is not unique. Infinitely many other equations of this general form represent the same line, for example,

$$\boldsymbol{x}(t) = \langle 2/3, -1, -1\rangle t + \langle -1, 3, 1\rangle$$

which uses the other of the two points as $\boldsymbol{x}_0$, and uses a $\boldsymbol{m}$ of a different length and in the opposite direction. The two things that all representations of this line will have in common, however, are that $\boldsymbol{x}_o$ will always correspond to a point of the line and that the differing choices for $\boldsymbol{m}$ will all be nonzero multiples of each other. Also, any line in $\mathbb{R}^3$ can be represented in this way, even a vertical line or one that parallels any coordinate axis.

1.2.3.2 Parametric form

While any line can be represented in the vector form described just above, this is not the only format to represent lines. Instead of using vectors, it is often useful to use parametric form which means using three scalar linear equations. Still the most important thing to keep in mind about these two formats is that they are equivalent.

Suppose a certain line $\mathcal{L}$ is given in vector form by $\boldsymbol{x}(t) = \boldsymbol{m}t + \boldsymbol{x}_o$ where $\boldsymbol{m} = \langle m_1, m_2, m_3\rangle$ and $\boldsymbol{x}_o = \langle a_o, b_o, c_o\rangle$. Then a parametric representation (*parametric form*)

---

4 Placing the scalar t after the vector $\boldsymbol{m}$ is opposite the normal order, but is done here simply to be more reminiscent of the scalar equation.

for this line $\mathcal{L}$ is given by the components of the vector equation:

$$x_1(t) = m_1 t + a_o$$
$$x_2(t) = m_2 t + b_o$$
$$x_3(t) = m_3 t + c_o$$

In this form, the parameter is t. As is the case with vector form, the representation in this form is *not* unique: a_o, b_o, and c_o correspond to any point on $\mathcal{L}$ and m_1, m_2, and m_3 correspond to the vector between any two points on the plane.

Example 1.7. Find an equation in parametric form for the line passing through the points $(1, 0, -2)$ and $(-1, 3, 1)$. (These are the same points, and thus the same line as in the previous example.)

Answer. The same $\boldsymbol{x}_o$ and $\boldsymbol{m}$ as in the previous example can be used to write the equation in parametric form:

$$x_1(t) = -2t + 1$$
$$x_2(t) = 3t$$
$$x_3(t) = 3t - 2$$

Again this representation is not unique.

1.2.4 Projections

Suppose there is a plane Π in $\mathbb{R}^3$, and a vector $\boldsymbol{a}$ not in this plane (see Figure 1.9). How can one find the vector lying in the plane closest to the given vector? Alternatively, if we have two vectors, how can one find the vector that parallels the second vector while being closest to the first? This subsection addresses these questions; interestingly, both have essentially the same answer.

Let us first consider the projection of one vector onto another; the definition of this projection is based on some simple geometry as shown in Figure 1.8. Given two vectors $\boldsymbol{a}$ and $\boldsymbol{b}$, the *vector projection* of $\boldsymbol{a}$ onto $\boldsymbol{b}$ is defined as

$$\mathbf{proj}_{\boldsymbol{b}}(\boldsymbol{a}) := \frac{\boldsymbol{a} \cdot \boldsymbol{b}}{|\boldsymbol{b}|^2}\boldsymbol{b} \tag{1.2}$$

provide that the vector $\boldsymbol{b} \neq \mathbf{0}$. To justify this definition, suppose that the angle between $\boldsymbol{a}$ and $\boldsymbol{b}$ is denoted θ. Then the length of the projection should be

$$\begin{aligned} |\mathbf{proj}_{\boldsymbol{b}}(\boldsymbol{a})| &= |\boldsymbol{a}| \cos\theta \\ &= \frac{|\boldsymbol{a}||\boldsymbol{b}| \cos\theta}{|\boldsymbol{b}|} \\ &= \frac{\boldsymbol{a} \cdot \boldsymbol{b}}{|\boldsymbol{b}|}. \end{aligned}$$

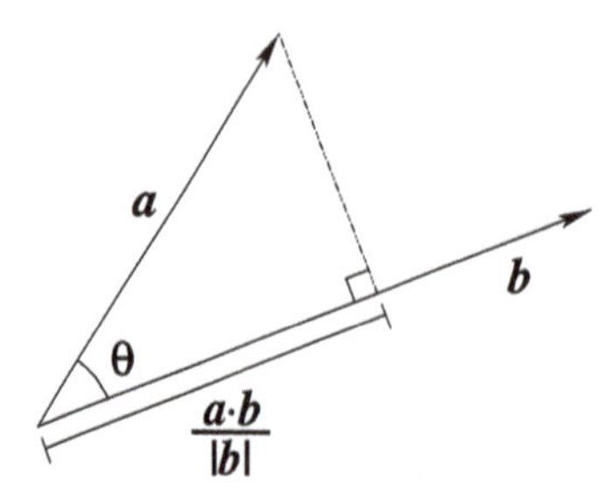

Figure 1.8: The length of the vector projection. This length is the distance along the vector $\boldsymbol{b}$ defined by the dashed perpendicular segment from the head of $\boldsymbol{a}$. The vector projection is then this length multiplied by the unit vector in the direction of $\boldsymbol{b}$.

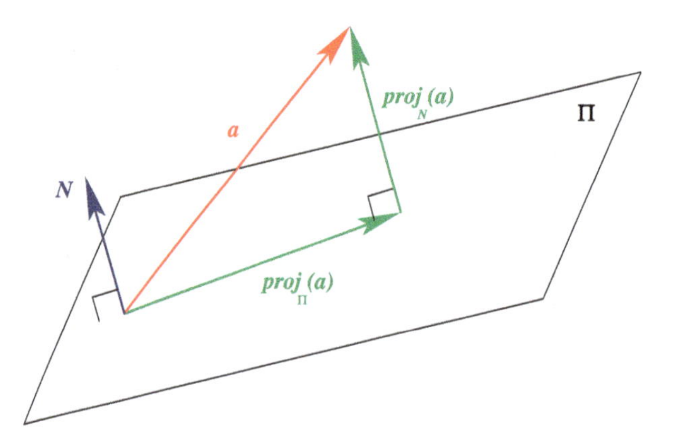

Figure 1.9: Projection of a vector $\boldsymbol{a}$ onto a plane Π.

The direction for $\mathbf{proj}_{\boldsymbol{b}}(\boldsymbol{a})$ is then given by the unit vector in the direction of $\boldsymbol{b}$, which of course is just $\boldsymbol{b}/|\boldsymbol{b}|$. Combining these length and direction results, one finds the vector projection given in (1.2), that is,

$$\mathbf{proj}_{\boldsymbol{b}}(\boldsymbol{a}) = \frac{\boldsymbol{a} \cdot \boldsymbol{b}}{|\boldsymbol{b}|^2}\boldsymbol{b} = \left(\frac{\boldsymbol{a} \cdot \boldsymbol{b}}{|\boldsymbol{b}|}\right)\left(\frac{\boldsymbol{b}}{|\boldsymbol{b}|}\right).$$

The *projection* of a vector onto a plane is depicted in Figure 1.9. The formal definition for this projection onto a plane Π is based on the observation that vector $\boldsymbol{a}$ should be the sum of its vector projection onto the normal vector $\boldsymbol{N}$ for the plane and its projection onto the corresponding plane:

$$\mathbf{proj}_{\Pi}(\boldsymbol{a}) := \boldsymbol{a} - \mathbf{proj}_{\boldsymbol{N}}(\boldsymbol{a})$$

Example 1.8. Find the vector in Π, the plane $2x - y + 5z = 1$, that is closest to the vector $\boldsymbol{a} = \langle -1, 0, 4\rangle$. Also find the projection of $\boldsymbol{a}$ onto any vector perpendicular to Π.

Answer. Notice that for this plane, an easy choice for a normal vector is $\boldsymbol{N} = \langle 2, -1, 5 \rangle$, and since this normal vector and $\boldsymbol{a}$ are not perpendicular, $\boldsymbol{a}$ is not in the plane Π. So what vector in this plane is closest to $\langle -1, 0, 4 \rangle$? Of course, it is the projection:

$$\begin{aligned} \mathbf{proj}_{\Pi}(\boldsymbol{a}) &= \langle -1, 0, 4 \rangle - \frac{\langle -1, 0, 4 \rangle \cdot \langle 2, -1, 5 \rangle}{2^2 + (-1)^2 + 5^2} \langle 2, -1, 5 \rangle \\ &= \langle -11/5, 3/5, 1 \rangle \end{aligned}$$

Notice that this vector *is* perpendicular to $\boldsymbol{N}$, and hence it does lie in the plane. On the other hand, the projection of $\boldsymbol{a}$ onto $\boldsymbol{N}$ (or any other vector perpendicular to Π) is

$$\mathbf{proj}_{\boldsymbol{N}}(\boldsymbol{a}) = \frac{\langle -1, 0, 4 \rangle \cdot \langle 2, -1, 5 \rangle}{2^2 + (-1)^2 + 5^2} \langle 2, -1, 5 \rangle = \langle 6/5, -3/5, 3 \rangle ,$$

and $\boldsymbol{a}$ is in fact the sum of these two projections.

1.3 Basic surfaces in $\mathbb{R}^3$

We have already seen one type of surface in $\mathbb{R}^3$: planes. This section discusses more general surfaces, particularly quadratic (or quadric) surfaces—those with only first and second degree terms.

Typically (generically), a surface in $\mathbb{R}^3$ is defined by a single equation. Equations for surfaces fall into two broad categories: explicit and implicit. If one can solve for z (or perhaps one of the other variables) in terms of the remaining two variables, then the equation for the surface is said to be *explicit*: $z = f(x, y)$ for some function $f : D \subset \mathbb{R}^2 \to \mathbb{R}$ where D is the domain of f. If one cannot solve for any of the three variables, then the equation of the surface is said to be *implicit*: $F(x, y, z) = 0$ for some function $F : D \subset \mathbb{R}^3 \to \mathbb{R}$. Here D is the domain of F, and 0 must be in the range. The most basic example of a surface whose equation can only be written implicitly in rectangular coordinates is a sphere: $x^2 + y^2 + z^2 = R^2$ (a sphere centered at the origin with radius R). Also notice that any surface that can be written explicitly as $z = f(x, y)$ can also be written implicitly using $F(x, y, z) = z - f(x, y)$.

1.3.1 Quadratic surfaces

Although it is easy to overstate their importance, there are a number of named quadratic surfaces in $\mathbb{R}^3$. The most significant of these will be introduced through a series of examples.

Example 1.9. Both the simplest and most important quadratic surface is the *unit sphere*:

$$x^2 + y^2 + z^2 = 1$$

This is of course the set of all points in $\mathbb{R}^3$ a distance 1 from the origin. In terms of the definition for a surface given above, $F(x,y,z) = x^2 + y^2 + z^2 - 1$ is one possible function that would define this surface. The general equation of a *sphere* in $\mathbb{R}^3$ is $(x - x_o)^2 + (y - y_o)^2 + (z - z_o)^2 = r^2$; this sphere is centered at (x_o, y_o, z_o) with radius r.

Example 1.10. The surface

$$\frac{x^2}{a^2} + \frac{y^2}{b^2} + \frac{z^2}{c^2} = 1$$

is an *ellipsoid* centered at the origin (see Figure 1.10). Here, $a, b, c > 0$ define the three principal axes; notice that $-a < x < a$, $-b < y < b$, and $-c < z < c$. When $a = b = c$, the ellipsoid becomes a sphere. Ellipsoids and spheres are the only bounded standard quadratic surfaces.

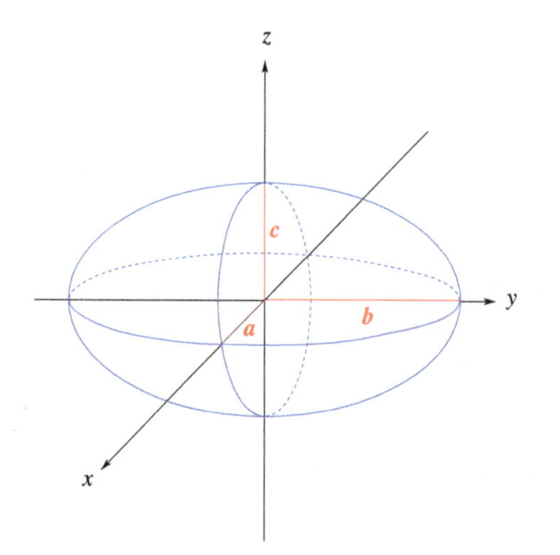

Figure 1.10: An ellipsoid centered at the origin. The three principal axes a, b, and c are shown in red. This shape is similar to that of a rugby ball; the ends of an American football are too pointed for it to be considered an ellipsoid.

Example 1.11. Consider the surface $2x^2 + 3y^2 - 6y - z + 8 = 0$. Because there are both quadratic and linear terms in y, one must complete the square with respect to y to find the standard form for this surface: $z = 2x^2 + 3(y - 1)^2 + 5$. This surface is an *elliptic paraboloid* opening upward in z with vertex at $(0, 1, 5)$. It is elliptic in that the scaling differs for the two quadratic terms. Thus the intersection of this surface with any horizontal plane $z = c$ (where $c > 5$ is a constant) is an ellipse.

Example 1.12. The surface $z^2 = x^2 + y^2$ is a *circular cone*. For any fix $r > 0$, the intersection of this cone with either of the planes $z = -r$ and $z = r$ is the circle $x^2 + y^2 = r^2$.

The origin is the vertex of this cone; this is the only singular point where there is no normal vector to the surface.

Example 1.13. A *circular cylinder* is a surface of the form $x_1^2 + x_2^2 = r^2$ in $\mathbb{R}^3$. Its axis is the third coordinate axis, here the x_3 axis. In general, the axis of the cylinder is the coordinate axis for the coordinate not mentioned in the equation.

Example 1.14. The final quadric surface mentioned here is a *hyperbolic paraboloid*:

$$z = \frac{x^2}{a^2} - \frac{y^2}{b^2}$$

This is the simplest example of a more general type of surface called a saddle surface; such surfaces will consider later. Saddle surfaces rise as one moves in one direction from the origin (here the x-direction), but fall as one moves in another direction (here the y-direction).

1.4 Polar, cylindrical, and spherical coordinates

There is an old aphorism that says "You can't put a square peg in a round hole." (Or is it "You can't put a round peg in a square hole.") This aphorism gets at a key issue in mathematics and science: some two- or three-dimensional problems are most easily described in rectangular coordinates, (either x and y in two dimensions, or x, y, and z in three dimensions), while others are most easily described by the distance between some object and a reference point (usually the origin) and some set of angles giving the direction of that object from that reference point.[5] The latter situation usually leads to polar coordinates in two-dimensional space, and either cylindrical and spherical coordinates in three-dimensional space. The relationship between these coordinate systems is discussed in this section.

1.4.1 Polar coordinates in $\mathbb{R}^2$

Consider a point on the standard x, y-plane ($\mathbb{R}^2$) as is shown in Figure 1.11. As is customary, suppose that position on this plane is measured by the two coordinate axes, with the x-axis on the horizontal and the y-axis on the vertical. By convention, the four quadrants defined by these axes are enumerated starting with the first quadrant where both x and y are positive, and increasing from there as one moves counterclockwise between the quadrants, again as shown in Figure 1.11. It is relatively easy to see that

5 Rectangular coordinates are also called Cartesian coordinates in honor of René Descartes (1596–1650).

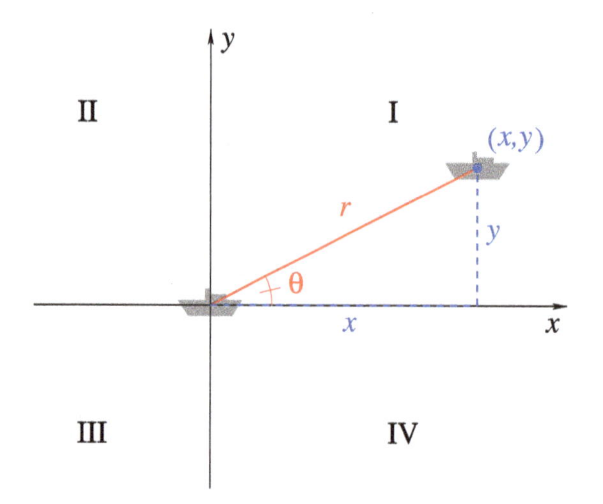

Figure 1.11: The standard x, y-plane, with one point distinguished. In rectangular coordinates, this point is (x, y). In polar coordinates, this point is (r, θ), a distance r from the origin at an angle θ with the positive x-axis. The four quadrants are enumerated by Roman numerals.

each point in the plane is uniquely determined by its corresponding x and y values, but there are other ways to locate points in the plane. The most important of these is polar coordinates.

To understand the motivation behind polar coordinates, it is perhaps helpful to think of ourselves as being on a ship on the surface of an ocean. The ocean surface is for our purposes a plane. Suppose we have a compass, so we can determine east, west, north, and south, and we can define east as the positive x direction and north as the positive y direction. Our ship then is at $(0, 0)$, the origin as in Figure 1.11. Suppose there is another ship within sight of our ship; how can we determine its location? There is no easy way to directly measure this other ship's x or y coordinate value, but it is easy to measure angle that a line between the two ships makes with the positive x direction (i. e., east). There are also a number of ways to measure the distance between the two ships, for example, by measuring the time between seeing a cannon flash and hearing the sound of that cannon. Notice that these two pieces of information, this angle and this distance, determine the position of the other ship on the ocean surface relative to the position of our ship.

Now let us return to the standard plane in Figure 1.11. Let r (for radial) be the distance between the point at (x, y) and the origin $(0, 0)$, and let θ be the angle between the positive x-axis and a ray (half-line) from the origin through the point at (x, y). This distance, angle, and ray, along with the point and the axes are depicted in red in Figure 1.11. Notice that generally by convention $r \geq 0$ and $\theta \in [0, 2\pi)$.

The relationship between rectangular and polar coordinates is given by right-angle trigonometry: Notice that r is the length of the hypotenuse of a right triangle one of whose sides has length x and the other has length y, and θ is the angle between the side of length x and the hypotenuse. Applying right-angle trigonometry, one finds that $x = r\cos\theta$ and $y = r\sin\theta$. To find r in terms of x and y, one can appeal to the Pythagorean theorem: $r^2 = x^2 + y^2$, and since r is a distance and therefore never negative, $r = \sqrt{x^2 + y^2}$. Finally, dividing the expressions for y and x and solving for θ yields $\theta = \mathrm{Tan}^{-1}(y/x)$, provided that $x \neq 0$.

What should always be remembered about the relationship between polar and rectangular coordinates? Along with the memorable image in Figure 1.11, three of the above equations are particularly simple and always worth remembering:

$$x = r\cos\theta$$
$$y = r\sin\theta$$
$$r^2 = x^2 + y^2$$

These three and Figure 1.11 can always be used to construct any of the other details.

Example 1.15.
(a) Which (x, y) value corresponds to $(r, \theta) = (6, \pi/6)$?
(b) Which (r, θ) value corresponds to $(x, y) = (-3, 4)$?

Answer. For (a), the solution is simply $x = 6\cos(\pi/6) = 3\sqrt{3}$ and $y = 6\sin(\pi/6) = 3$. For (b), $r = \sqrt{(-3)^2 + 4^2} = 5$, while $\tan(\theta) = y/x = -4/3 \Rightarrow \theta \simeq 2.214$. There are in fact two values of θ for which $\tan(\theta) = -4/3$, one in the second quadrant, and one in the fourth. The correct choice must match the location of (x, y), which in this case is in the second quadrant.

Example 1.16. Please describe the curve whose equation in polar coordinates is $r = 4\cos\theta$.

Answer. At first glance, one might be tempted to simply replace r and θ by their expressions in rectangular coordinates. This approach, however, does not produce an equation that is easy to recognize:

$$\sqrt{x^2 + y^2} = 4\cos(\mathrm{Tan}^{-1}(y/x))$$

While this expression could be reduced to the desired result, it is perhaps better to go back to the polar expression and simplify it. Specifically, if $r = 4\cos\theta$, then multiplying both sides by r produces $r^2 = 4r\cos\theta$ which in rectangular coordinates is just $x^2 + y^2 = 4x$. Bringing the right-hand side to the left and completing the square, one finds that $(x-2)^2 + y^2 = 4$. This expression is now easy to recognize and describe as the equation of the circle of radius 2 centered on the point $(x, y) = (2, 0)$.

1.4.2 Cylindrical and spherical coordinates in $\mathbb{R}^3$

Polar coordinates in $\mathbb{R}^2$ generalize in two particularly important ways in $\mathbb{R}^3$. The first is simply to take two of the three rectangular coordinates for $\mathbb{R}^3$ and replace them by polar coordinates. The result is *cylindrical coordinates*: (r, θ, z) where as in polar coordinates, $x = r\cos\theta$ and $y = r\sin\theta$. The third coordinate z is unchanged between rectangular and cylindrical coordinates. The relationship between the rectangular and cylindrical coordinates for a point in $\mathbb{R}^3$ is shown graphically in Figure 1.12.

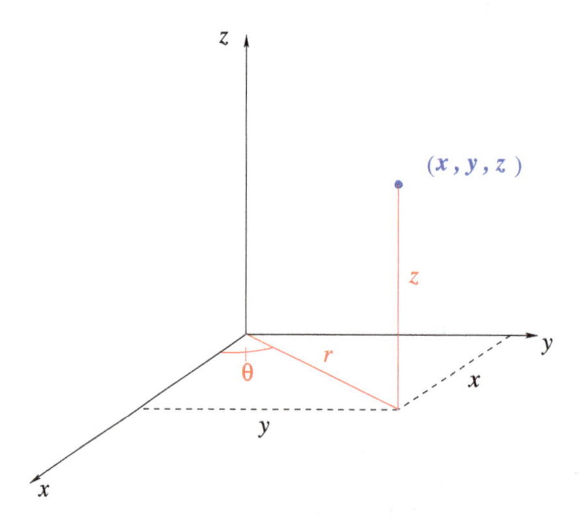

Figure 1.12: Cylindrical coordinates (r, θ, z) along with rectangular coordinates (x, y, z) for $\mathbb{R}^3$.

Example 1.17. Consider the surface $z = x^4 - y^4$; what is its equations in cylindrical coordinates?

Answer. Notice that

$$\begin{aligned} x^4 - y^4 &= (x^2 - y^2)(x^2 + y^2) \\ &= r^2(\cos^2\theta - \sin^2\theta)r^2 \\ &= r^4\cos(2\theta), \end{aligned}$$

so in cylindrical coordinates the equation for this surface is $z = r^4 \cos 2\theta$ which is perhaps a bit simpler than the rectangular version. Notice that the power on x, y, and r is preserved under this conversion in coordinates—this is typically the case when one starts with a polynomial in rectangular coordinates.

If cylindrical coordinates is a straightforward extension to three-dimensional space of two-dimensional polar coordinates, then the second extension, spherical coordinates, is a much more spiritually faithful extension. The definitions in *spherical coordinates* in $\mathbb{R}^3$ are based on the same key idea as polar coordinates are in $\mathbb{R}^2$. Suppose one is standing at the origin in $\mathbb{R}^3$; consider the position of a point that in rectangular coordinates is located at (x, y, z). What is the distance between the origin and this point? As in polar coordinates, the answer comes from the Pythagorean theorem and gives us the first spherical coordinate: $\rho := \sqrt{x^2 + y^2 + z^2}$. Once this distance is determined, as in the polar case, the rest of the work is down to setting up the correct angles. But whereas one angle was needed in $\mathbb{R}^2$, two angles are needed in $\mathbb{R}^3$. Fortunately, the first of these is just the polar or cylindrical angle θ. Here, this angle is defined by projecting the ray from the origin to the point at (x, y, z) onto the x, y-plane. The result of this projection is the ray in our polar coordinate discussion above, and θ is again the angle this ray in the x, y-plane makes with the positive x-axis. The second angle (the third coordinate) in the spherical triple is ϕ, which is defined relative to the positive z-axis: the angle ϕ is measured from the positive z-axis to the ray in $\mathbb{R}^3$ from the origin to the point at (x, y, z). This third coordinate then must take on values between 0 (at the positive z-axis) and π (at the negative z-axis). In the middle, when $\varphi = \pi/2$, is the x, y-plane.[6] The relationship between the rectangular, cylindrical, and spherical coordinates for a point in $\mathbb{R}^3$ is shown graphically in Figure 1.13.

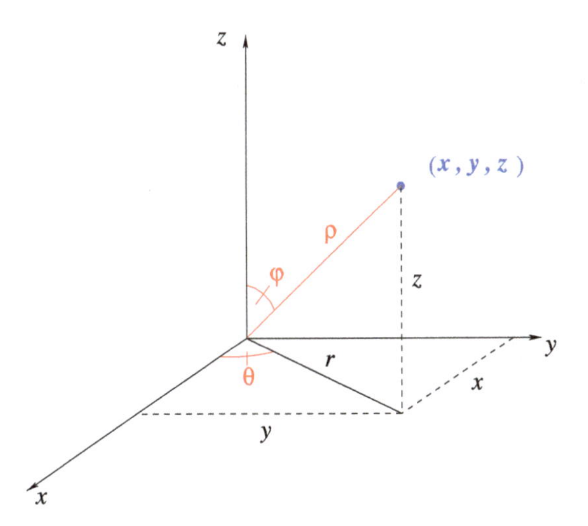

Figure 1.13: Spherical coordinates (ρ, θ, ϕ) along with rectangular coordinates (x, y, z), and cylindrical coordinates (r, θ, z) for $\mathbb{R}^3$.

6 Unlike cylindrical coordinates where the coordinate-triple is almost always given as (r, θ, z), the order, names, and even definitions of the three spherical coordinates varies among authors and disciplines. The angle φ sometimes takes on values in $[-\pi/2, \pi/2]$, rather than $[0, \pi]$, with $\varphi = 0$ being the x, y-plane. Also many authors use ρ to denote density, and hence must use something else for the spherical radial distance. And the order that the angles θ and φ appear in the coordinate triple may be reversed. Nevertheless, this text will stick to the definitions and order given above since they are traditional in calculus.

Example 1.18. If $(x, y, z) = (1, 2, -3)$, please represent this point in spherical coordinates.

Answer. Here, $\rho = \sqrt{1^2 + 2^2 + (-3)^2} = \sqrt{14}$, $\theta = \text{Tan}^{-1}(2/1) \simeq 1.107$ (about 63 degrees), and $\phi = \text{Cos}^{-1}(z/\rho) = \text{Cos}^{-1}(-3/\sqrt{14}) \simeq 2.501$ (about 143 degrees).

Example 1.19. Describe in cylindrical and spherical coordinates the intersection of the sphere $x^2 + y^2 + z^2 = 9$ and the half-cone $z^2 = x^2 + y^2$, $z > 0$.

Answer. To obtain the cylindrical representation of this intersection, combine the equations for the sphere and the cone, eliminating both x and y. One finds that $2z^2 = 9 \Rightarrow z = 3\sqrt{2}/2$ (since $z > 0$), and thus $r^2 = x^2 + y^2 = z^2 \Rightarrow r = 3\sqrt{2}/2$ as well. Since θ can take on any value, the intersection in cylindrical coordinates is $(r, \theta, z) = (3\sqrt{2}/2, \theta, 3\sqrt{2}/2)$ where $0 \le \theta < 2\pi$. So this is the circle with radius $3\sqrt{2}/2$ centered on the z-axis in the plane $z = 3\sqrt{2}/2$.

In spherical coordinates, the entire half-cone is simply $z^2 = r^2$, which reduces to $z = r$ because both z and r are positive. Thus

$$1 = \frac{z}{r} = \frac{\rho \sin \phi}{\rho \cos \phi} = \tan \phi$$

which reduces to $\phi = \pi/4$. For the spherical radius, the intersection is $\rho^2 = r^2 + z^2 = 9 \Rightarrow \rho = 3$. Hence intersecting circle is $(\rho, \theta, \phi) = (3, \theta, \pi/4)$ where again, $0 \le \theta < 2\pi$ This circle is shown in Figure 1.14.

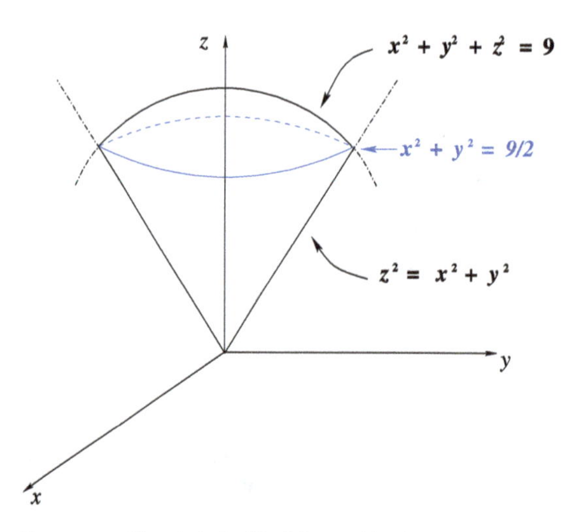

Figure 1.14: The circle (in blue) formed by the intersection of the sphere $x^2 + y^2 + z^2 = 9$ and the half-cone $z^2 = x^2 + y^2$, $z > 0$. The equations for this circle in rectangular coordinates is $x^2 + y^2 = 9/2$, $z = 3\sqrt{2}/2$.

Exercises 1

1.1. If $\boldsymbol{a} = \langle 1, 3, -2 \rangle$ and $\boldsymbol{b} = \langle 4, -1, -1 \rangle$, please compute the length of $\boldsymbol{a}$, the length of $\boldsymbol{b}$, the sum of these two vectors, and their dot and cross products.

Answer. $|\boldsymbol{a}| = \sqrt{14}$, $|\boldsymbol{b}| = 3\sqrt{2}$, $\boldsymbol{a} + \boldsymbol{b} = \langle 5, 2, -3 \rangle$, $\boldsymbol{a} \cdot \boldsymbol{b} = 3$, $\boldsymbol{a} \times \boldsymbol{b} = \langle -5, -7, -13 \rangle$.

1.2. If $\boldsymbol{a} = \langle 5, -1, 8 \rangle$ and $\boldsymbol{b} = \langle x, y, z \rangle$, what equation must be satisfied if $\boldsymbol{a} \perp \boldsymbol{b}$?

1.3. What is the (smaller) angle θ between the vectors $\langle 1, 3, -2 \rangle$ and $\langle 4, -1, -1 \rangle$?

Answer. $\theta = \mathrm{Cos}^{-1}(\sqrt{7}/14) \simeq 1.381$

1.4. Consider the six vectors $\boldsymbol{a}_1 = \langle 1, 0, -3 \rangle$, $\boldsymbol{a}_2 = \langle -2, 1, 4 \rangle$, $\boldsymbol{a}_3 = \langle \pi, 0, -3\pi \rangle$, $\boldsymbol{a}_4 = \langle 3, 2, 1 \rangle$, $\boldsymbol{a}_5 = \langle 1, -1/2, -2 \rangle$, and $\boldsymbol{a}_6 = \langle 1, 8, 0 \rangle$. Which of these vectors are perpendicular to each other? Which are parallel?

1.5. What must be true if, given two vectors $\boldsymbol{a}$ and $\boldsymbol{b}$, $(\boldsymbol{a} + \boldsymbol{b}) \perp (\boldsymbol{a} - \boldsymbol{b})$?

1.6. Please write the vector $\mathbf{v} = \langle 3, -2, 5 \rangle$ as the product of its length and a unit vector that gives its direction.

1.7. If $\boldsymbol{a}, \boldsymbol{b} \in \mathbb{R}^3$, solve $(\boldsymbol{a} - c\boldsymbol{b}) \cdot \boldsymbol{b} = 0$ for a scalar $c \in \mathbb{R}$ in terms of $\boldsymbol{a}$ and $\boldsymbol{b}$. Is there always an c, no matter how the vectors $\boldsymbol{a}$ and $\boldsymbol{b}$ are chosen? Is c unique?

1.8. Use the definitions of the dot and cross products to prove the first three identities in Proposition 1.

1.9. Please show that for any vectors $\boldsymbol{a}, \boldsymbol{b} \in \mathbb{R}^3$, the cross product $\boldsymbol{a} \times \boldsymbol{b}$ is perpendicular to both $\boldsymbol{a}$ and $\boldsymbol{b}$. Hint: Use the definition of $\boldsymbol{a} \times \boldsymbol{b}$ and compute the dot products $\boldsymbol{a} \times \boldsymbol{b} \cdot \boldsymbol{a}$ and $\boldsymbol{a} \times \boldsymbol{b} \cdot \boldsymbol{b}$. Notice that the notation $\boldsymbol{a} \times \boldsymbol{b} \cdot \boldsymbol{a}$ implicitly requires that one compute the cross product first and then the dot product.

1.10. Prove that $|\boldsymbol{a} \times \boldsymbol{b}| = |\boldsymbol{a}||\boldsymbol{b}| \sin\theta$. Hint: Compute $|\boldsymbol{a} \times \boldsymbol{b}|^2$ from the definition, then separately expand out $|\boldsymbol{a}|^2|\boldsymbol{b}|^2 \sin^2\theta = |\boldsymbol{a}|^2|\boldsymbol{b}|^2(1 - \cos^2\theta) = |\boldsymbol{a}|^2|\boldsymbol{b}|^2 - (\boldsymbol{a} \cdot \boldsymbol{b})^2$ to arrive at the same polynomial.

1.11. Please prove the *triangle inequality*: If $\boldsymbol{a}, \boldsymbol{b} \in \mathbb{R}^3$, then $|\boldsymbol{a} + \boldsymbol{b}| \leq |\boldsymbol{a}| + |\boldsymbol{b}|$. Hint: Write $|\boldsymbol{a} + \boldsymbol{b}|^2$ as the dot product of a vector with itself, then distribute out this product, use the Cauchy–Schwarz inequality to bound $\boldsymbol{a} \cdot \boldsymbol{b}$ and form $(|\boldsymbol{a}| + |\boldsymbol{b}|)^2$.

1.12. What is an equation for the plane passing through the point $(4, 0, -2)$ perpendicular to the vector $\langle -1, 5, 3 \rangle$?

Answer. $x - 5y - 3z = 10$ (or any equivalent equation).

1.13. In Example 1.5, choose the vectors $\boldsymbol{a}$ and $\boldsymbol{b}$ in a different way, and choose a different point on the plane for (x_o, y_o, z_o), then carry out the computation to arrive at the same equation for the plane as in the example.

1.14. Find an equation for the plane passing through the points $(2, 1, 0)$, $(1, 0, 2)$, and $(0, 2, 1)$.

1.15. Please find a vector equation for the line passing through the points $(1, -1, 3)$ and $(-3, 2, 1)$.

Answer. $\mathbf{x}(t) = \langle 1, -1, 3\rangle + \langle 4, -3, 2\rangle\, t$ (there are many other possibilities).

1.16. Please find a parametric representation for the line of intersection for the planes $x + 2y - 3z = 4$ and $3x - y + z = -1$.

1.17. What equation describes all of the points that are equidistant from both the point $(-1, 0, 2)$ and the point $(2, 3, -1)$.

Answer. $2x + 2y - 2z = 3$

1.18. Please find the vector projection of $\langle 6, 5, -1\rangle$ onto $\langle -2, 1, -3\rangle$.

Answer. Here, $\boldsymbol{a} = \langle 6, 5, -1\rangle$, $\boldsymbol{b} = \langle -2, 1, -3\rangle$, and $\mathbf{proj}_{\boldsymbol{b}}(\boldsymbol{a}) = \langle 4, -2, 6\rangle/7$.

1.19. For the plane Π: $-3x - 2y + z = 5$, and the vector $\boldsymbol{a} = \langle 1, -1, 3\rangle$, please find $\mathbf{proj}_{\boldsymbol{N}}(\boldsymbol{a})$ and $\mathbf{proj}_{\Pi}(\boldsymbol{a})$.

1.20. Consider the vector form for the equation of a line: $\mathbf{x}(t) = \mathbf{x}_o + \boldsymbol{m}t$ where $\mathbf{x} = \langle x, y, z\rangle$, $\mathbf{x}_o = \langle x_o, y_o, z_o\rangle$, and $\boldsymbol{m} = \langle m_1, m_2, m_3\rangle$.

(a) By eliminating the time parameter t, obtain the following system of two linear equations:

$$\frac{x - x_o}{m_1} = \frac{y - y_o}{m_2} = \frac{z - z_o}{m_3}$$

provided that $m_i \neq 0$ for $i = 1, 2, 3$.

(b) These two equations determine a line as the intersection of two planes. What is the equation for each of the plane? (The choice of the two equations is not unique.)

1.21. If $z = g(x, y) = \sqrt{3 - x^2 - y^2}$, how can one describe this surface in words? What is an equivalent implicit form for the equation for this surface? Why is $x^2 + y^2 + z^2 - 3 = 0$ not an equivalent implicit form?

1.22. Is it possible to give a single explicit equation for the surface given implicitly by $x^2 - y + z^2 - 1 = 0$? If possible, please write one example of an equivalent explicit equation and name the surface.

1.23. Please describe the following surfaces as accurately as possible:

(a) $x^2 + y^2 + z = 0$

(b) $x^2 - 6x + 5y^2 + 10y - 63 + z^2 = 0$

(c) $5 - x^2 + y^2 - z = 0$

(d) $x^2 + y^2 + z^2 + 1 = 0$

Answer. (a) A paraboloid opening downward with vertex at the origin $(0, 0, 0)$.

1.24. For $(x, y) = (1, 2) \in \mathbb{R}^2$, what are the corresponding polar coordinates?

Answer. $(r, \theta) = (\sqrt{5}, \mathrm{Tan}^{-1}(2)) \simeq (2.236, 1.107)$.

1.25. Please write $(x, y, z) = (-1, 3, -7) \in \mathbb{R}^3$ in both cylindrical and spherical coordinates.

Answer. $(r, \theta, z) = (\sqrt{10}, \mathrm{Tan}^{-1}(-3), -7) \simeq (3.162, -1.249, -7)$; $(\rho, \theta, \phi) = (\sqrt{59}, \mathrm{Tan}^{-1}(-3), \mathrm{Cos}^{-1}(-7/\sqrt{59})) \simeq (7.681, -1.249, 2.717)$.

1.26. Please express the vector from the origin to (x, y, z) first in cylindrical, then in spherical coordinates. What are the cylindrical and spherical expressions of this vector for the specific point $(\sqrt{3}, 1, 0)$?

1.27. Please transform the equations of each of the following surfaces from rectangular coordinates to both cylindrical and spherical coordinates (where possible, also name the surface):

(a) $z = x^2 + y^2$
(b) $z = x^2 - y^2$
(c) $z^2 = x^2 + y^2$
(d) $x = y$
(e) $z = x^2$
(f) $x = y$
(g) $(x - 5)^2 + (y - 12)^2 = 169$
(h) $z^2 = x^2 + y^2$

1.28. Please transform the equations of each of the following surfaces from cylindrical coordinates to rectangular coordinates (where possible, also name the surface):

(a) $r = z/\cos\theta$
(b) $z = x^2 - y^2$
(c) $r^2 = z^2$
(d) $x = y$
(e) $z = x^2$
(f) $x = y$
(g) $x = y$
(h) $r = 6\cos\theta - 8\sin\theta$

2 Vector functions

2.1 Limits, derivatives, and integrals for vector functions

We now turn our attention to extending calculus from scalar functions (as one studies in a basic calculus course) to vector functions. This first section introduces and extends the most important calculus concepts to the realm of vectors. Later sections discuss and interpret vector functions more geometrically.

Definition. A *vector function* is a function $\boldsymbol{v}$ defined on $\mathbb{R}$ or some subset $D \subset \mathbb{R}$ whose outcomes are vectors in $\mathbb{R}^n$. The set D on which $\boldsymbol{v}$ is defined is the *domain* and the set of all outcomes (vectors) is the *range*. In symbols, this definition is $\boldsymbol{v} : D \subset \mathbb{R} \to \mathbb{R}^n$.

Thus for any $t \in D$, our vector function results in a vector in $\mathbb{R}^n$:

$$\boldsymbol{v}(t) = \langle v_1(t), v_2(t), \ldots, v_n(t) \rangle$$

As was the case earlier in dealing with constant vectors, this component representation for vector functions is frequently used in our discussion.[1]

Example 2.1. A simple example of a vector function is

$$\boldsymbol{v}(t) = \left\langle 3t, \frac{1}{1+t^2}, \sqrt{1+t^2} \right\rangle .$$

Here, the domain is $D = \mathbb{R}$ because there are no restrictions on the choice of t, and the range is a subset of $\mathbb{R}^3$ because this vector function has 3 components. The first component runs through all real numbers, but the second must be in the interval $(0, 1]$, and the third is always greater than or equal to 1. The vectors $\boldsymbol{v}(0)$ and $\boldsymbol{v}(1)$ are shown in Figure 2.1. A second example is

$$\boldsymbol{u}(t) = \left\langle \sqrt{1-t^2}, \frac{1}{1-t^2} \right\rangle .$$

For this $\boldsymbol{u}$, the range is a subset of $\mathbb{R}^2$, while the domain is $D = (-1, 1)$. This domain is the intersection of the domains of each of the two components: $\text{Domain}(\sqrt{1-t^2}) = [-1, 1]$ while $\text{Domain}(1/(1-t^2)) = (-\infty, -1) \cup (-1, 1) \cup (1, \infty)$.

There are of course many, many other vector functions. Can you think of one that takes on values in $\mathbb{R}^3$?

Definition. The *limit* of a vector function can also be defined componentwise:

$$\lim_{t \to t_o} \boldsymbol{v}(t) := \left\langle \lim_{t \to t_o} v_1(t), \lim_{t \to t_o} v_2(t), \ldots, \lim_{t \to t_o} v_n(t) \right\rangle$$

provided that each and every component limit on the right exists and is finite.

1 We use t or τ as the independent or domain variable because this variable often represents *time*.

https://doi.org/10.1515/9783110660609-002

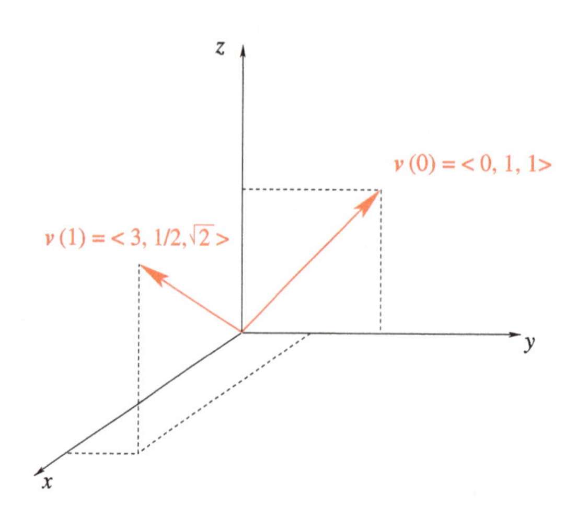

Figure 2.1: Two specimen vectors $\boldsymbol{v}(0)$ and $\boldsymbol{v}(1)$ for the vector function $\boldsymbol{v}(t) = \langle 3t, 1/(1+t^2), \sqrt{1+t^2}\rangle$. The vectors are drawn emanating from the origin:.

Example 2.2. For $\boldsymbol{v}$ given in the previous example,

$$\lim_{t\to t_o} \boldsymbol{v}(t) = \left\langle \lim_{t\to t_o} 3t, \lim_{t\to t_o} \sqrt{1+t^2}, \lim_{t\to t_o} \frac{1}{1+t^2} \right\rangle = \left\langle 3t_o, \sqrt{1+t_o^2}, \frac{1}{1+t_o^2} \right\rangle.$$

Notice that solution is valid for all $t_o \in \mathbb{R}$ since each component is continuous on its entire domain. Another example: Suppose that

$$\boldsymbol{w}(t) = \left\langle \frac{\cos^2(1-t)-1}{(1-t)^2}, \frac{1-t}{1-\sqrt{t}} \right\rangle$$

for $t \neq 1$. Then

$$\lim_{t\to 1} \boldsymbol{w}(t) = \left\langle \lim_{t\to 1} \frac{\cos^2(1-t)-1}{(1-t)^2}, \lim_{t\to 1} \frac{1-t}{1-\sqrt{t}} \right\rangle = \langle -1, 2\rangle.$$

In computing the limits of the components, one has all the usual tools: trig identities, factoring, l'Hôpital's rule, etc. Thus in the previous example, one can used l'Hôpital's rule to compute the limit of the first component and can notice that $1-t = (1-\sqrt{t})(1+\sqrt{t})$ to compute the limit of the second component.

Remark. It is possible to give a ϵ-δ definition for the limit of a vector function as is normally done for scalar limits. Indeed in a mathematical sense, ϵ-δ definition should be preferred. But the definition given here is equivalent, and it is easier to use computationally.

As in the single-variable case, the velocity of the particle is the derivative of position, and acceleration is the derivative of velocity. So to continue our discussion, we need to define the derivative of a vector function; this definition must be consistent with the single-variable case and uses the definition of limit above.

Definition. The derivative of a vector function can now be defined as a vector function limit: For a vector function $\boldsymbol{v}$, the *derivative* is

$$\begin{aligned}\dot{\boldsymbol{v}}(t) \equiv \frac{d\boldsymbol{v}}{dt} &:= \lim_{h\to 0} \frac{\boldsymbol{v}(t+h) - \boldsymbol{v}(t)}{h} \\ &= \left\langle \lim_{h\to 0} \frac{v_1(t+h) - v_1(t)}{h}, \lim_{h\to 0} \frac{v_2(t+h) - v_2(t)}{h}, \dots, \lim_{h\to 0} \frac{v_n(t+h) - v_n(t)}{h} \right\rangle \\ &= \langle \dot{v}_1(t), \dot{v}_2(t), \dots, \dot{v}_n(t) \rangle\end{aligned}$$

provided that each component limit (derivative) exists.

Example 2.3. Suppose that $\boldsymbol{v}(t) := \langle \cos t, \sin t, 1\rangle$. Here, the natural domain is again $D = \mathbb{R}$ (t can take on any real value). Computing componentwise, one finds that $\dot{\boldsymbol{v}}(t) = \langle -\sin t, \cos t, 0\rangle$. Notice that in this example, $\boldsymbol{v}(t) \perp \dot{\boldsymbol{v}}(t)$ (i. e., $\boldsymbol{v}(t)\cdot\dot{\boldsymbol{v}}(t) = 0$) regardless of the value of t. This situation is a special case of a general result that will be discussed below.

Remark. The use of a dot over a time-dependent variable to denote derivative goes back to Isaac Newton and is similar to prime notation. Here, both dot and prime notation will be used, with dot being reserved for differentiation with respect to time, and prime indicating differentiation with respect to some other (specified) variable.

Finally, we closed this section by defining the integral of a vector function; the definition is again given componentwise.

Definition. For a vector function $\boldsymbol{v}$, its *integral* is a vector function $\boldsymbol{V}$ defined as

$$\begin{aligned}\boldsymbol{V}(t) &\equiv \int_{t_o}^{t} \boldsymbol{v}(\tau)\, d\tau \\ &:= \left\langle \int_{t_o}^{t} v_1(\tau)\, d\tau, \int_{t_o}^{t} v_2(\tau)\, d\tau, \dots, \int_{t_o}^{t} v_n(\tau)\, d\tau \right\rangle\end{aligned}$$

provided that each of the component integrals exist. The time t_o is some convenient reference time; often $t_o = 0$.

Notice that when each of the components v_i is continuous, then by the fundamental theorem of calculus,

$$\dot{\boldsymbol{V}}(t) = \boldsymbol{v}(t) = \langle v_1(t), v_2(t), \dots, v_n(t)\rangle\ ,$$

that is, the derivative of a vector function define as a function of the upper limit of integration of an integral is simply the vector integrand evaluated at that variable of differentiation t.

Example 2.4. As in the previous example, suppose that $\boldsymbol{v}(t) := \langle \cos t, \sin t, 1\rangle$. Then with $t_o = 0$,

$$\begin{aligned} \boldsymbol{V}(t) &\equiv \int_0^t \boldsymbol{v}(\tau)\, d\tau \\ &= \left\langle \int_0^t \cos\tau\, d\tau, \int_0^t \sin\tau\, d\tau, \int_0^t 1\, d\tau \right\rangle \\ &= \langle \sin t, 1 - \cos t, t\rangle . \end{aligned}$$

Also

$$\dot{\boldsymbol{V}}(t) = \langle \cos t, \sin t, 1\rangle = \boldsymbol{v}(t) .$$

2.2 Parametric Curves in $\mathbb{R}^2$ and $\mathbb{R}^3$

This section is a brief pause in our discussion of vector functions to turn our attention to a very related topic: parametric curves.

Definition. Suppose that f, g, and h are three continuous, real-valued functions, all defined on some interval $I \subset \mathbb{R}$. In symbols, $f, g, h : I \subset \mathbb{R} \to \mathbb{R}$. For these functions and this interval I, the *parametric curve* C is the set of all points (x, y, z) of the form $x = f(t)$, $y = g(t)$, $z = h(t)$ for some $t \in I$. Again, in symbols $C := \{(x, y, z) \in \mathbb{R}^3 \,|\, x = f(t), y = g(t), z = h(t) \text{ for some } t \in I\}$. Also $\gamma := (f, g, h)$ is a *parameterization* of C, and one can write $\gamma(t)$ for the point $(f(t), g(t), h(t))$ that lies on the curve C and is reached when the parameter takes on the value t, or for the singleton set $\{(x, y, z) \in \mathbb{R}^3 \,|\, x = f(t), y = g(t), z = h(t)\}$. Indeed one can think of the curve C being traced out by the parameterization γ as the parameter t runs through all its values.

A generic example of a parametric curve is shown in Figure 2.2. If $I = [a, b]$ is a finite, closed interval for some real numbers $a < b$ (in general, it could be infinite), then γ begins at the point $\alpha = (f(a), g(a), h(a))$ and ends at the point $\omega = (f(b), g(b), h(b))$. To obtain a parametric curve in $\mathbb{R}^2$ (the x, y-plane), simply leave off the third function h and the third variable z: $\gamma := \{(x, y) \in \mathbb{R}^2 \,|\, x = f(t), y = g(t) \text{ for some } t \in I\}$.

Example 2.5. Suppose that $I = [-2, 1]$, that $f(t) = 2t + 1$ and $g(t) = (1 + t)^2$. Then the parametric curve C is a parabola with $x = 2t + 1$ and $y = (1 + t)^2$, beginning at $\alpha = (-3, 1)$ and ending at $\omega = (3, 4)$. This curve C is shown in Figure 2.3. Because C is in the x, y-plane, and because one can solve for t in terms of x, a single equation for the curve can found by eliminating the parameter t. Notice that $t = (x - 1)/2$ and this expression for t can be substituted into g: $y = (1+t)^2 = (1+(x-1)/2)^2 = (x+1)^2/4$. Also notice that one cannot solve for t or x in terms of y.

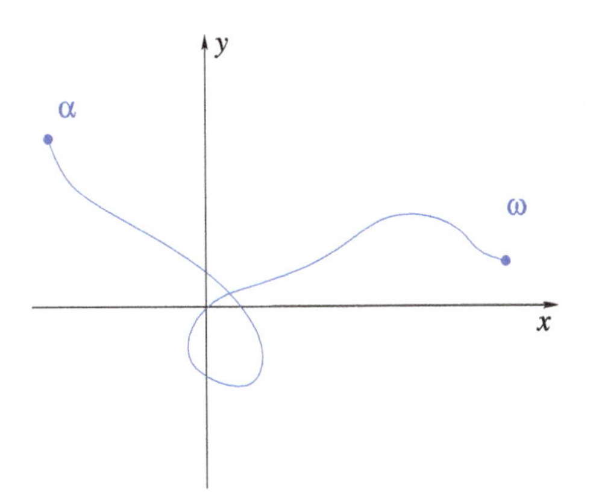

Figure 2.2: A parametric curve for some finite interval in the x, y-plane. The curve crosses itself, so $\gamma(t_1) = \gamma(t_2)$ for some $t_1 < t_2$.

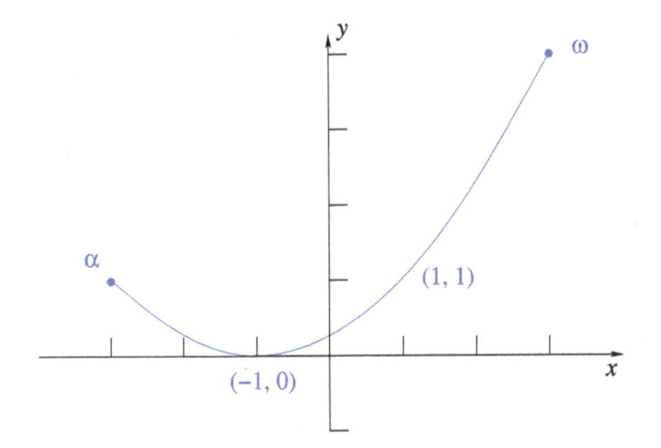

Figure 2.3: Parabolic arc from $\alpha = (-3, 1)$ to $\omega = (3, 4)$ along $y = (x + 1)^2/4$.

For a given interval I and given functions f, g, and h, there is a single (unique) curve C. Interestingly, though, this does not work the other way around: a given curve C will have many parameterizations using different I, f, g, and/or h. Which parameterization is preferred often depends on what one is trying to do.

Example 2.6. Suppose that $I_1 = [-1, 1]$, that $f_1(t) = 2t$ and $g_1(t) = t^2$. Also suppose that $I_2 = [-2, 2]$, that $f_2(t) = t$, and $g_2(t) = t^2/4$. Both of these parameterizations describe the curve $y = x^2/4$ beginning at $(-2, 1)$ and ending at $(2, 1)$.

The previous example may seem rather trivial, but there are more important ones.

Example 2.7. Let C be the portion of the unit circle lying in the first quadrant: $C = \{(x, y)|x^2 + y^2 = 1, x \geq 0, y \geq 0\}$. This circular arc is depicted in Figure 2.4. One parameterization is simply $f_1(t) = \sqrt{1 - t^2}$, $g_1(t) = t$ and $I_1 = [0, 1]$ beginning at $(1, 0)$ and ends at $(0, 1)$. Another rather different parameterization, also beginning at $(1, 0)$ and ending

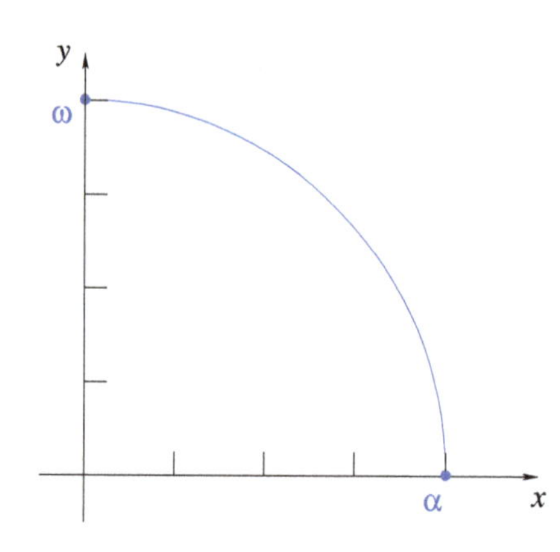

Figure 2.4: Circular arc from $\alpha = (1, 0)$ to $\omega = (0, 1)$ along $1 = x^2 + y^2$.

at $(0, 1)$, is $f_2(t) = \cos t$, $g_2(t) = \sin t$ and $I_2 = [0, \pi/2]$. The first parameterization may seem simpler, and it may be preferred in some circumstances, but it does not extend to the entire circle. The second does; one simply needs to extend I_2 to $I_2 = [0, 2\pi]$.

The real advantage of a parametric representation for a curve becomes most clear in $\mathbb{R}^3$ for more complicated and thus more interesting curves, or curves that cross themselves.

Example 2.8. Let $I = [0, 4]$, with $f(t) = \cos(\pi t)$, $g(t) = \sin(\pi t)$ and $h(t) = t$. This parametric curve is two loops of a helix: $C = \{(x, y, z) | x = \cos(\pi t),\ y = \sin(\pi t),\ z = t,\ t \in [0, 4]\}$ and is shown in Figure 2.5. It would be difficult to describe this helix accurately without using a parametric representation.

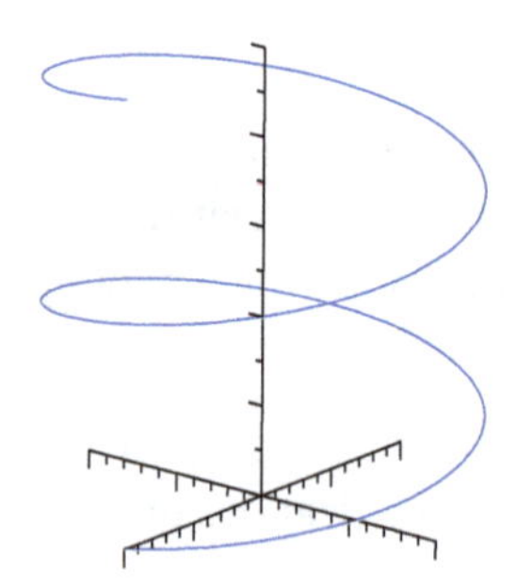

Figure 2.5: The portion of the helix $C = \{(x, y, z) | x = \cos(\pi t),\ y = \sin(\pi t),\ z = t,\ t \in [0, 4]\}$.

Example 2.9. Let $I = (-\infty, \infty)$, with $f(t) = 1 - t^2$ and $g(t) = t(1 - t^2)$. Then this parametric curve is an alpha curve: $C = \{(x, y) | x = 1 - t^2,\ y = t(1 - t^2),\ t \in (-\infty, +\infty)\}$. It is possible to give a single equation in x and y that represents this curve: $y^2 = (1-x)x^2$ but notice that one can not solve for either x or y. So the parametric representation again

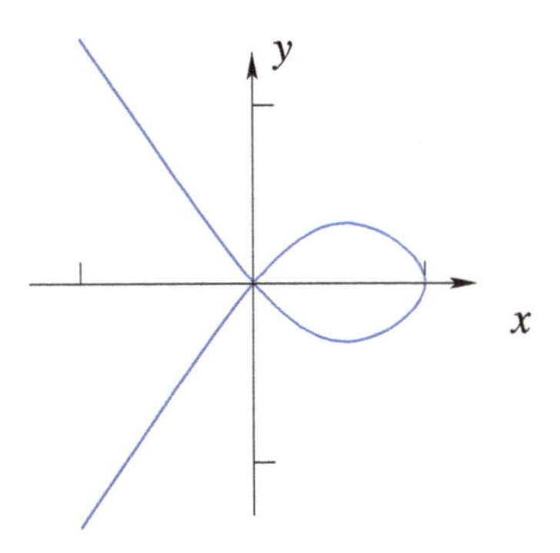

Figure 2.6: The alpha curve $y^2 = (1-x)x^2$.

has advantages. This alpha curve is shown in Figure 2.6; the curve is in the shape of the Greek letter alpha.

Notice that the alpha curve in the previous example (Example 2.9) and the generic curve shown in Figure 2.2 both have the feature that they cross over themselves, that is, there are two or more distinct times t_1 and t_2 such that $\gamma(t_1) = \gamma(t_2)$. This is an important issue to watch for. Another is whether or not a curve is smooth, a term that is defined next.

Definition. A parametric curve is *smooth* if there is a parameterization $\gamma = (f, g, h)$ defined on some interval $I \subset \mathbb{R}$ where (1) each function f, g and h is continuously differentiable on I (meaning that f, g, and h are all differentiable and f', g', and h' are all continuous), and (2) there is no $t \in I$ such that $f'(t) = g'(t) = h'(t) = 0$.

Anyone who has studied calculus should not be surprised that the definition of "smooth" involves differentiability, but the second condition (that there be no time t where all the derivatives are simultaneously zero) may be a surprise. Why is zero a problem here? The next example should make this clear.

Example 2.10. Let $I = [-1, 1]$ and $\gamma(t) = (f(t), g(t)) = (t^3, t^2)$. Since both f and g are both monomials, they are both differentiable, and one might think that nothing too interesting will happen. But a plot of this curve shows otherwise: Figure 2.7. If the parameter is eliminated, and we solve for y in terms of x, the resulting equations is $y = x^{2/3}$. This curve has a sharp point at $(x, y) = (0, 0)$ known as a *cusp*.

Understanding what goes wrong when all the derivatives are simultaneously zero is perhaps best seen by considering a particle moving along the curve whose position on the plane at time t is $(f(t), g(t))$ (a view that will be explored more fully in the next section). If all of the derivatives are simultaneously zero, the particle will stop at least for a moment. This means that when it starts again, it can move in a very different direction from the one it had been heading before. This is why there can be a sharp corner or cusp at such a point. Interestingly, there is a very important example of this

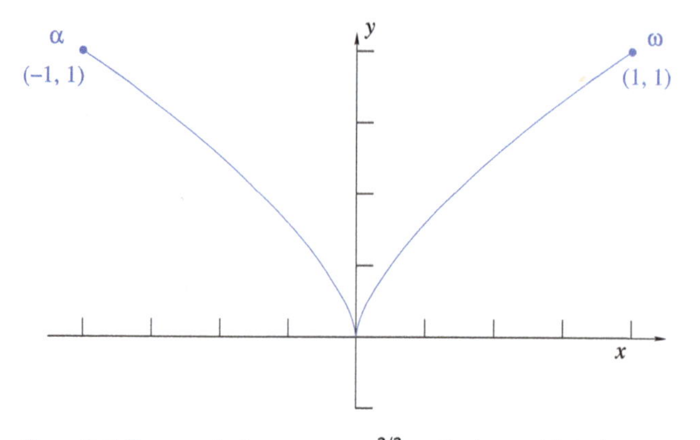

Figure 2.7: The cusp in the curve $y = x^{2/3}$ on the interval $[-1, 1]$.

sort of thing in weather forecasting: hurricanes. Predicting where a hurricane is heading is much easier when the hurricane is moving; when it stops, it can be difficult or impossible to say where it will go next after it starts moving again.

2.3 Particle motion in $\mathbb{R}^2$ and $\mathbb{R}^3$

Now we connect the two previous sections and use them to represent particle motion in two and three dimensions. There are many situations that involve particle motion, but one of the most dramatic was a news item literally around the world on October 4, 1957: Спутник (Sputnik). This was the first artificial satellite ever to orbit the earth; it was more-or-less spherical, a little more than half a meter in diameter, and it admitted regular radio beep that made it easy to track. But how could someone describe its position, velocity, acceleration, etc.? As one might expect, the central mathematical concepts for this description are vector functions and parametric curves.

Suppose that the position of Sputnik or any moving particle is determined by a vector function $\boldsymbol{x}(t) = \langle x_1(t), x_2(t), x_3(t) \rangle$ whose tail is at some fixed reference point and whose head is at the position of the particle at time t (see Figure 2.8). The particle's path is then the parametric curve traced out of the particle, parameterized by the

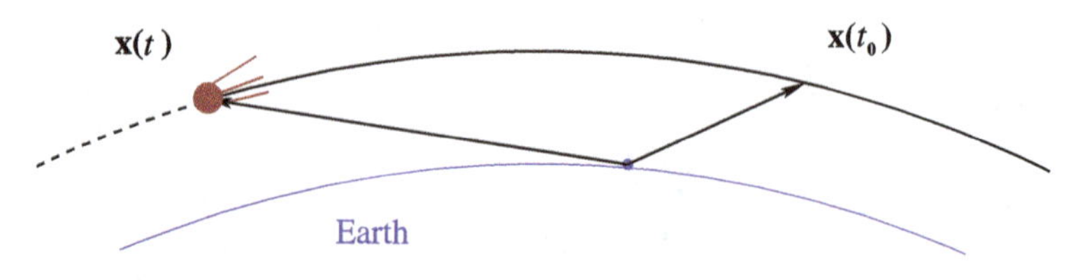

Figure 2.8: Sputnik orbiting the earth, tracked by a vector function $\boldsymbol{x}$. Its position at time t is $\boldsymbol{x}(t)$, and its initial position at some reference time t_o is $\boldsymbol{x}(t_o)$. The blue point on the earth is the location of the tracking station.

components of the vector function:

$$f(t) = x_1(t), \quad g(t) = x_2(t), \quad h(t) = x_3(t)$$

As in single-variable calculus, velocity of the particle is defined to be the derivative of position, and acceleration is defined to be the derivative of velocity. A key question then is "Where do the velocity and acceleration vectors lie relative to the path of the particle and the position vector?" The rest of this section is devoted to answering this question.

2.3.1 Tangent vectors

How is a vector function related to its derivative? To answer this question, one needs to recall the definition of the derivative for vector functions: For the vector function $\boldsymbol{x}$, by definition its *derivative* is

$$\dot{\boldsymbol{x}}(t) := \lim_{h \to 0} \frac{\boldsymbol{x}(t+h) - \boldsymbol{x}(t)}{h}$$

provided this limit exists. What can be seen from this definition is that the derivative is the limit of a difference quotient whose numerator is a secant vector connecting points on the particle path (see Figure 2.9). As $h \to 0$, the direction of this secant vector converges to a direction tangent to the particle path. The length of this secant vector converges to zero, but of course, the denominator also goes to zero. As a result, the length of the derivative vector (which is the speed that the particle is moving along its path) is a real value between zero (if the numerator goes to zero faster than the denominator) and infinity (if the denominator goes to zero faster than the numerator). All of this makes clear that the direction of the derivative is tangent to the path traced out by the particle or its position vector, and it motivates the following definitions.

Definition. Suppose that $\boldsymbol{x}(t)$ is the position vector (as a function of time t) of a particle moving in either $\mathbb{R}^2$ or $\mathbb{R}^3$. Then $\boldsymbol{v}(t) := \dot{\boldsymbol{x}}(t)$ is the *velocity vector*, and $\boldsymbol{a}(t) := \dot{\boldsymbol{v}}(t) = \ddot{\boldsymbol{x}}(t)$ is the *acceleration vector*. The length of the velocity vector is the *speed*[2] that the particle moves along its path: $\dot{s}(t) := |\boldsymbol{v}(t)|$. Notice that $\forall\, t,\ 0 \leq \dot{s}(t) < +\infty$.

Since velocity is always tangent to the curve traced out by its position vector, the velocity vector can be used to define a tangent vector with unit length, provided the velocity is nonzero.

2 One might have expected that s would be speed, rather than $\dot{s}$, but in fact s is traditionally arc length, and this will be discussed below.

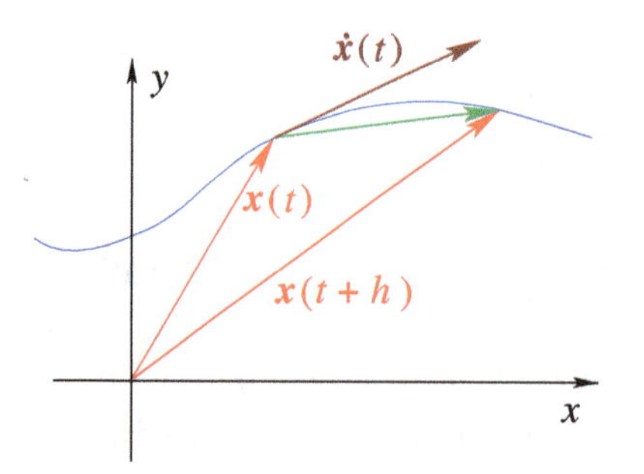

Figure 2.9: Secant vectors $\boldsymbol{x}(t+h) - \boldsymbol{x}(t)$ (in green) moves to the tangent vector $\dot{\boldsymbol{x}}(t)$ (in brown) as $h \to 0$. As $h \to 0$, the head of the green secant vector moves backward along the blue curve toward the head of $\boldsymbol{x}(t)$ (in red). Since the derivative is the limit of the difference quotient, the direction of the derivative must be the limit of the direction of the numerator. This limiting direction is tangent to the curve.

Definition. The *unit tangent vector* $\boldsymbol{T}$ is defined as

$$\boldsymbol{T}(t) := \frac{\boldsymbol{v}(t)}{|\boldsymbol{v}(t)|}$$

provided that the velocity vector $\boldsymbol{v}(t) \neq \boldsymbol{0}$.

An illustrative example would seem to now be in order.

Example 2.11. Consider a vector function that traces out an elliptical helix:

$$\boldsymbol{x}(t) = \langle \cos(\pi t), 2\sin(\pi t), t \rangle$$

for $0 \leq t \leq 7$. Please find a parametric representation of the curve traced out by this function, the speed, and the velocity, acceleration, and unit tangent vectors.

Answer. A parameterization for the curve traced out by $\boldsymbol{x}(t)$ is given by simply using the components of $\boldsymbol{x}$ as f, g, and h: the parametric curve is $y = \{(x, y, z) \in \mathbb{R}^3 \mid x = \cos(\pi t), y = 2\sin(\pi t), z = t \text{ for some } t \in [0, 7]\}$. If one thinks of a particle moving along the helix whose position is given by $\boldsymbol{x}(t)$, for $t \in [0, 7]$, then the velocity of the particle is given by

$$\boldsymbol{v}(t) = \dot{\boldsymbol{x}}(t) = \langle -\pi \sin(\pi t), 2\pi \cos(\pi t), 1 \rangle ,$$

the acceleration is

$$\boldsymbol{a}(t) = \dot{\boldsymbol{v}}(t) = \ddot{\boldsymbol{x}}(t) = \left\langle -\pi^2 \cos(\pi t), -2\pi^2 \sin(\pi t), 0 \right\rangle$$

and speed is $\dot{s}(t) = \sqrt{\pi^2 \sin^2(\pi t) + 4\pi^2 \cos^2(\pi t) + 1} = \sqrt{1 + \pi^2 + 3\pi^2 \cos^2(\pi t)}$. Finally, the unit tangent vector is

$$\boldsymbol{T}(t) = \frac{\langle -\pi \sin(\pi t), 2\pi \cos(\pi t), 1 \rangle}{\sqrt{1 + \pi^2 + 3\pi^2 \cos^2(\pi t)}}.$$

2.3.2 Normal vectors

So far, we have found that the derivative of a vector function is tangent to the curve traced out by that vector function and that this derivative can be used to find the unit tangent vector $\boldsymbol{T}$. Interestingly, this unit tangent vector can be used to find a unit normal vector $\boldsymbol{N}$ that is perpendicular to the particle path (in this sense, "normal" means "perpendicular"). The key mathematical result that allows us to define $\boldsymbol{N}$ is the following proposition.

Proposition 2. *Suppose that* $\forall\,\tau$, $\boldsymbol{u}(\tau)$ *is a unit vector:* $|\boldsymbol{u}(\tau)| = 1\,\forall\,\tau$. *(Here, the independent variable or parameter* τ *may denote time or may be some other quantity.) Then* $\forall\,\tau$

$$\boldsymbol{u}(\tau) \perp \dot{\boldsymbol{u}}(\tau)$$

provided that $\dot{\boldsymbol{u}}(\tau)$ *exists and* $\dot{\boldsymbol{u}}(\tau) \neq \mathbf{0}$.

Proof. Since $|\boldsymbol{u}(\tau)| = 1$, then $\forall\,\tau$, $|\boldsymbol{u}(\tau)|^2 = \boldsymbol{u}(\tau)\cdot\boldsymbol{u}(\tau) = 1$. Differentiating with respect to τ, one finds that

$$\frac{d}{d\tau}(\boldsymbol{u}(\tau)\cdot\boldsymbol{u}(\tau)) = \frac{d}{d\tau}(1) = 0.$$

But by the product rule, this equation becomes $\forall\tau$,

$$\dot{\boldsymbol{u}}(\tau)\cdot\boldsymbol{u}(\tau) + \boldsymbol{u}(\tau)\cdot\dot{\boldsymbol{u}}(\tau) = 2\boldsymbol{u}(\tau)\cdot\dot{\boldsymbol{u}}(\tau) = 0,$$

which implies that $\boldsymbol{u}(\tau) \perp \dot{\boldsymbol{u}}(\tau)$. □

Thus since $\boldsymbol{T}(t)$ is a unit vector $\forall\,t$, Proposition 2 implies that $\dot{\boldsymbol{T}}(t)$ is perpendicular to $\boldsymbol{T}(t)$ and also to the curve traced out by our vector function $\boldsymbol{x}$ (see Figure 2.10). Hence $\dot{\boldsymbol{T}}(t)$ is a normal vector. All of this leads to the following definition.

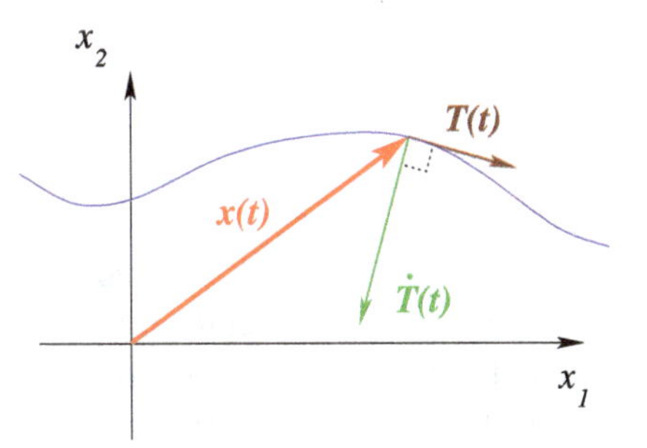

Figure 2.10: The unit tangent vector $\boldsymbol{T}(t)$ and its derivative $\dot{\boldsymbol{T}}(t)$ for the vector function $\boldsymbol{x}$.

Definition. The *unit normal vector* $\boldsymbol{N}$ is defined as

$$\boldsymbol{N}(t) := \frac{\dot{\boldsymbol{T}}(t)}{|\dot{\boldsymbol{T}}(t)|}$$

provided that $\dot{\boldsymbol{T}}(t)$ exists and $\dot{\boldsymbol{T}}(t) \neq \boldsymbol{0}$.

Remark. One might think that the derivative of a unit vector is itself a unit vector; this is not the case. Therefore, one must divide $\dot{\boldsymbol{T}}(t)$ by its length to obtain a unit vector.

Now we can continue our previous example.

Example 2.11. Consider again the elliptical helix traced out by

$$\boldsymbol{x}(t) = \langle \cos(\pi t), 2\sin(\pi t), t \rangle .$$

Then as before, the unit tangent vector is

$$\boldsymbol{T}(t) = \frac{\langle -\pi \sin(\pi t), 2\pi \cos(\pi t), 1 \rangle}{\sqrt{1 + \pi^2 + 3\pi^2 \cos^2(\pi t)}}$$

and one can compute $\dot{\boldsymbol{T}}(t)$ directly:

$$\dot{\boldsymbol{T}}(t) = \frac{-\pi^2 \langle (1 + 4\pi^2)\cos(\pi t),\ 2(1 + \pi^2)\sin(\pi t),\ 3\pi^3 \cos(\pi t)\sin(\pi t) \rangle}{(1 + \pi^2 + 3\pi^2 \cos^2(\pi t))^{3/2}} .$$

But notice that since we are only interested in the direction of $\dot{\boldsymbol{T}}(t)$, not its length, we may drop the scalar coefficient (including the denominator) and simply use the vector to compute $\boldsymbol{N}(t)$:

$$\boldsymbol{N}(t) = \frac{\langle (1 + 4\pi^2)\cos(\pi t),\ 2(1 + \pi^2)\sin(\pi t),\ 3\pi^3 \cos(\pi t)\sin(\pi t) \rangle}{\sqrt{1 + 8\pi^2 + 4\pi^4 + 3\sin^2(\pi t) + 12\pi^4 \cos^2(\pi t) + 9\pi^6 \cos(\pi t)\sin(\pi t)}} .$$

Still this result is rather messy; see how much simpler these calculations are for a circular helix (Exercise 2.9).

2.3.3 Acceleration

Unlike the velocity vector, which is always tangent to the particle path, there is no fixed direction for the acceleration vector. But for a particle moving in either $\mathbb{R}^2$ or $\mathbb{R}^3$, it is possible to write the acceleration vector as the sum to two vectors whose directions and lengths can be interpreted. Specifically, provided that $|\boldsymbol{v}(t)| \neq 0$ and that all of the derivatives make sense, acceleration can be written as

$$\boldsymbol{a}(t) = \frac{d}{dt}(\boldsymbol{v}(t)) = \frac{d}{dt}\left(|\boldsymbol{v}(t)| \frac{\boldsymbol{v}(t)}{|\boldsymbol{v}(t)|} \right)$$

$$
\begin{aligned}
&= \frac{d}{dt}(\dot{s}(t)\,\boldsymbol{T}(t)) \\
&= \ddot{s}(t)\,\boldsymbol{T}(t) + \dot{s}(t)\,\dot{\boldsymbol{T}}(t).
\end{aligned}
$$

So acceleration can be written as the sum of two terms, the first being *tangent* to the particle path, the second, *normal* to that path. The coefficient of the unit tangent vector $\boldsymbol{T}$ is $\ddot{s}(t)$, the time derivative of speed. This means that $\ddot{s}(t)$ is the magnitude of the acceleration in the tangential direction. This is the sort of acceleration one causes in a car by pushing down on either the accelerator or brake pedals; it changes the speed, but not the direction of motion.

By Proposition 2 above, one can see that the second term is normal to the curve: since $\boldsymbol{T}(t)$ is a unit vector $\forall t$, $\dot{\boldsymbol{T}}(t) \perp \boldsymbol{T}(t)$. But understanding the meaning of the coefficient of $\dot{\boldsymbol{T}}(t)$ is not yet easy because although $\boldsymbol{T}(t)$ is a unit vector, $\dot{\boldsymbol{T}}(t)$ in general is *not*. Also since the particle path is independent of the speed at which the particle moves along the curve, using time t as the parameter for the motion is not best for understanding the meaning of the normal coefficient. To resolve these two issues, we need to introduce arc length, the distance the particle has moved along the curve, and then use arc length rather than time to locate points on the curve traced out by the vector function. Arc length is defined and discussed in the next section, and then we return to our discussion of acceleration.

2.4 Arc length

In the previous section, speed was symbolized by $\dot{s}(t)$; this symbolism is traditional, but also leads to the obvious question: "What is speed the derivative of ?", or in other words, "What is s?" The answer is arc length.

2.4.1 Arc length between fixed points α and ω

Definition. Suppose that a given curve C in three-dimensional space begins at a point α, ends at a point ω, and is traced out by a vector function $\boldsymbol{x}(t) = \langle x_1(t), x_2(t), x_3(t)\rangle$ with $\boldsymbol{x}(0)$ corresponding to α and $\boldsymbol{x}(1)$ corresponding to ω (see Figure 2.11). Suppose also that the components of $\boldsymbol{x}$ are each differentiable. Then *arc length* s is the distance along C from α to ω and is given by the integral

$$
s := \int_0^1 |\dot{\boldsymbol{x}}(t)|dt = \int_0^1 \sqrt{\left(\frac{dx_1}{dt}\right)^2 + \left(\frac{dx_2}{dt}\right)^2 + \left(\frac{dx_3}{dt}\right)^2}\,dt
$$

provided this integral exists. If the arc length s is finite, then the curve C is *rectifiable*.

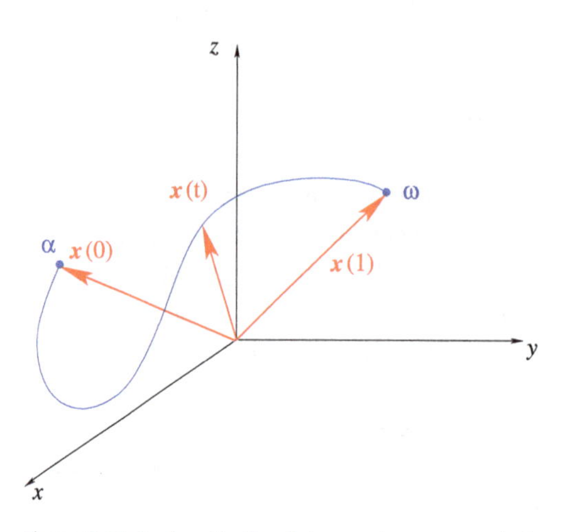

Figure 2.11: Arc length: the distance along a curve between two specific points, α and ω. The curve C is traced out by the vector function $\boldsymbol{x}$. Things associated with the curve C are in blue; those associated with the vector function $\boldsymbol{x}$ are in red.

If each of the component derivatives dx_1/dt, dx_2/dt, and dx_3/dt are continuous (i. e., if the components themselves are continuously differentiable), then the integral defining s will exist and be finite, so in this case, the curve C is rectifiable.

Remarks.

1. Arc length can be defined for curves traced out by vector functions that are *not* differentiable, but for our purposes, it is sufficient to only define it for differentiable vector functions. One can extend this definition in the obvious way for vector functions that are piecewise differentiable by computing the arc length of each piece.
2. Essentially, the same definition works in two-dimensional space. One must simply leave out the z-component of the vector function and the dz/dt term in the integrand.
3. One might expect that using $[0, 1]$ as the integration interval would be unnecessarily restrictive. This is not the case since for a given curve C, we can always rescale or translate the vector function so that $t = 0$ corresponds to α and $t = 1$ corresponds to ω. So this choice of $[0, 1]$ is simply a matter of convenience, not a restriction.

Why does the above integral represent distance along the curve? The answer can most easily be seen in Figure 2.12. There the integration interval $[0, 1]$ is partitioned into n subintervals, and the end points of these subintervals correspond to points on the curve C. One can now sum up the lengths of the line segments that connect successive points on C:

$$\sum_{i=1}^{n} \sqrt{(x_i - x_{i-1})^2 + (y_i - y_{i-1})^2 + (z_i - z_{i-1})^2}$$

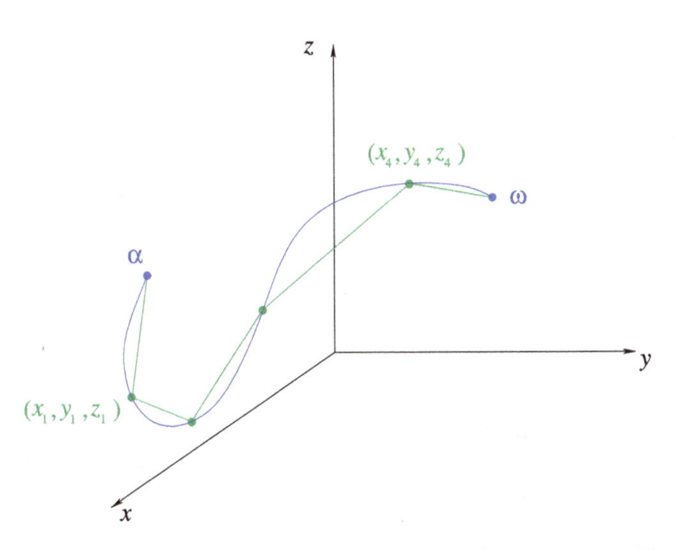

Figure 2.12: Arc length: the curve C is approximated by a sequence of line segments between two specific points, α and ω. Again things associated with the curve C are in blue; those associated with the approximating segments are in green.

$$= \sum_{i=1}^{n} \sqrt{\left(\frac{x_i - x_{i-1}}{t_i - t_{i-1}}\right)^2 + \left(\frac{y_i - y_{i-1}}{t_i - t_{i-1}}\right)^2 + \left(\frac{z_i - z_{i-1}}{t_i - t_{i-1}}\right)^2} (t_i - t_{i-1}) \,.$$

Taking the limit as the length of the longest subinterval in the partition (the *norm* of the partition) goes to zero (and thus n goes to infinity), one sees that fractions inside the square roots all approach derivatives, while the summation approaches an integral. The result is the expression for s given in the definition above.

Now let us consider several examples. The first is a simple example set in just the x, y-plane, the second is more interesting, and the third can only be computed numerically.

Example 2.12. Suppose that $\boldsymbol{x}(t) := \langle 3t/2, t^{3/2} \rangle$. Then $\dot{\boldsymbol{x}}(t) = \langle dx/dt, dy/dt \rangle = \langle 3/2, 3\sqrt{t}/2 \rangle$, $\alpha = (0, 0)$, $\omega = (3/2, 1)$, and

$$s = \int_0^1 |\dot{\boldsymbol{x}}(t)| dt = \frac{3}{2} \int_0^1 \sqrt{1 + t}\, dt = 2\sqrt{2} - 1 \,.$$

Example 2.13. Now suppose that $\boldsymbol{x}(t)$ traces out a portion of a circular helix: $\boldsymbol{x}(t) := \langle \cos 4\pi t, \sin 4\pi t, t \rangle$. As shown in Figure 2.13, as t runs from 0 to 1, $\boldsymbol{x}(t)$ traces out two cycles of our helix. But despite this being a more interesting curve, its arc length is computed exactly as in the previous example: $\dot{\boldsymbol{x}}(t) = \langle -4\pi \sin 4\pi t, 4\pi \cos 4\pi t, 1 \rangle$, $\alpha = (1, 0, 0)$, $\omega = (1, 0, 1)$, and

$$s = \int_0^1 |\dot{\boldsymbol{x}}(t)| dt = 4\pi \int_0^1 \sqrt{\sin^2 4\pi t + \cos^2 4\pi t + 1}\, dt = 4\pi\sqrt{2} \,.$$

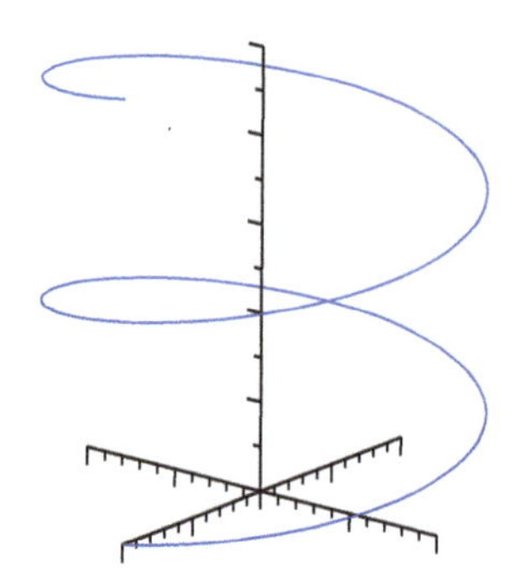

Figure 2.13: Two cycles of the helix whose arc length is computed in Example 2.13.

Example 2.14. Our final example in this set shows that not all smooth curves have arc length integrals that can be computed in closed form (integrated exactly). Suppose that $\boldsymbol{x}(t) := \langle t, t^3\rangle$. Then $\dot{\boldsymbol{x}}(t) = \langle 1, 3t^2\rangle$, $\alpha = (0,0)$, $\omega = (1,1)$, and

$$s = \int_0^1 |\dot{\boldsymbol{x}}(t)|dt = \int_0^1 \sqrt{1+9t^4}\,dt\,.$$

Unfortunately, this integral cannot be computed in closed form; the exact value of s this time can only be approximated numerically: $s \simeq 1.548$.

2.4.2 Arc length as a function of time: $s(t)$

Up to now, we have consider only the arc length of a curve between two fixed points; now we turn our attention to arc length as a function of time where the beginning point is still fixed, but the ending point now varies with t.

Definition. Let C be a curve in two-dimensional or three-dimensional space traced out by a vector function $\boldsymbol{x}(t)$ for $t \in \mathbb{R}$. Let $\boldsymbol{x}(0)$ be the vector from the origin to a reference point on C. Then

$$s(t) := \int_0^t |\dot{\boldsymbol{x}}(\tau)|d\tau$$

is *arc length* as a function of t. It is the distance along the curve C from the reference point to the point on the curve at the head of $\boldsymbol{x}(t)$ (see Figure 2.14) provided this integral exists.

Remark. Notice that when $t < 0$ then $s(t) < 0$, (arc length can be negative).

The above definition for $s(t)$ and the fundamental theorem of calculus give the answer to the question at the start of this section: By the fundamental theorem of calculus, if the integrand is continuous, the derivative of any integral that is a function

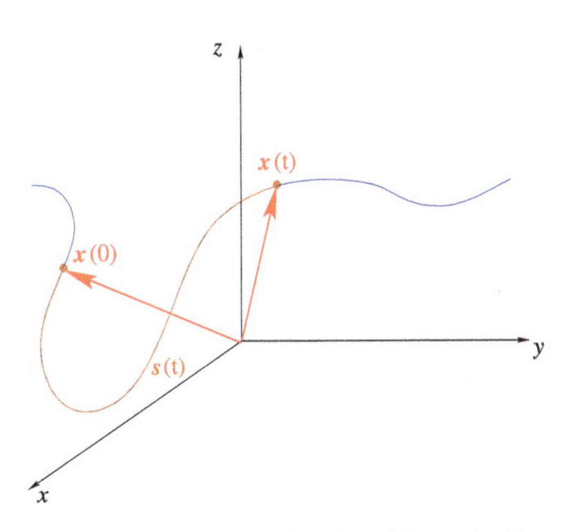

Figure 2.14: Arc length as a function of time t: the distance along a curve between some reference point corresponding to $\boldsymbol{x}(0)$ and a point at time t corresponding to $\boldsymbol{x}(t)$. Again, things associated with the curve C are in blue; those associated with the vector function $\boldsymbol{x}$ are in red. Things associated with arc length are in copper.

of the upper limit of integration is just that integrand evaluated at the upper limit of integration. So in this case

$$\dot{s}(t) = \frac{d}{dt}\left(\int_0^t |\dot{\boldsymbol{x}}(\tau)|d\tau\right) = |\dot{\boldsymbol{x}}(t)| = \dot{s}(t)\,.$$

The final equality above is just a reiteration of the fact that both $|\dot{\boldsymbol{x}}(t)|$ and $\dot{s}(t)$ were described as speed in the previous section.

Example 2.15. Suppose that a particle moves along a curve C with its velocity given by $\boldsymbol{v}(t) = \langle \cos t, \sin t\rangle$. Please find $s(t)$.

Answer. Recall that $\dot{s}(t) = |\boldsymbol{v}(t)|$; thus in this case, $\dot{s}(t) = \sqrt{\cos^2 t + \sin^2 t} = 1$, implying that $s(t) = t$. (Remember that $s(0) = 0$ because we measure the arc length from the point on the curve corresponding to $t = 0$.)

2.5 Acceleration decomposition

We can now complete our decomposition of acceleration. Recall that provided that $|\boldsymbol{v}(t)| \neq 0$ and that all of the derivatives make sense,

$$\begin{aligned}\boldsymbol{a}(t) &= \frac{d}{dt}(\boldsymbol{v}(t)) = \frac{d}{dt}\left(|\boldsymbol{v}(t)|\,\frac{\boldsymbol{v}(t)}{|\boldsymbol{v}(t)|}\right)\\ &= \frac{d}{dt}(\dot{s}(t)\,\boldsymbol{T}(t))\\ &= \ddot{s}(t)\,\boldsymbol{T}(t) + \dot{s}(t)\,\dot{\boldsymbol{T}}(t).\end{aligned}$$

Previously, we discussed the first term; now we consider the second. To understand the second term, the best parameter to measure position along the curve is arc length s. Using arc length allows one to describe the curve without regard to the speed of motion. By the chain rule,[3]

$$\dot{\boldsymbol{T}}(t) = \frac{d\boldsymbol{T}}{dt} = \frac{d\boldsymbol{T}}{ds}\frac{ds}{dt} = \boldsymbol{T}'(s(t))\dot{s}(t) \tag{2.1}$$

where prime ($'$) denotes differentiation with respect to s. One now forms a unit vector here in the same way as always: for a nonzero vector, divide the vector by its length:

$$\frac{d\boldsymbol{T}}{ds} = \boldsymbol{T}'(s) = |\boldsymbol{T}'(s)|\frac{\boldsymbol{T}'(s)}{|\boldsymbol{T}'(s)|} = |\boldsymbol{T}'(s)|\boldsymbol{n}(s)$$

where by definition $\boldsymbol{n}(s) := \frac{\boldsymbol{T}'(s)}{|\boldsymbol{T}'(s)|}$ is the *unit normal vector* as a function of arc length. Notice that since $|\boldsymbol{T}'(s)|$ is here always positive, $\boldsymbol{n}(s)$ points in the same normal direction as $\boldsymbol{T}'(s)$. This direction is into the direction that the motion is changing, that is, into the curve. Also notice that $\boldsymbol{n}(s(t)) = \boldsymbol{N}(t)$.

Now combining both of the above steps, one finds that the acceleration can be written as

$$\begin{aligned}
\boldsymbol{a}(t) &= \ddot{s}(t)\,\boldsymbol{T}(t) + \dot{s}(t)\,\dot{\boldsymbol{T}}(t) \\
&= \ddot{s}(t)\,\boldsymbol{T}(t) + (\dot{s}(t))^2\,\boldsymbol{T}'(s(t)) \\
&= \ddot{s}(t)\,\boldsymbol{T}(t) + (\dot{s}(t))^2|\boldsymbol{T}'(s(t))|\,\boldsymbol{n}(s(t)) \\
&= \ddot{s}(t)\,\boldsymbol{T}(t) + (\dot{s}(t))^2\kappa(s(t))\,\boldsymbol{n}(s(t)) \\
&= a_T(t)\,\boldsymbol{T}(t) + a_N(t)\,\boldsymbol{n}(s(t))
\end{aligned}$$

where $\kappa(s) := |\boldsymbol{T}'(s)|$ is defined to be the *curvature* of the motion, $a_T(t) := \ddot{s}(t)$ is the (scalar) tangential component of the acceleration, and $a_N(t) := (\dot{s}(t))^2\kappa(s)$ is the (scalar) normal component (see Figure 2.15). This curvature $\kappa(s)$ measures the tendency of the motion to turn (or curve) *into* the motion of the particle, that is, into the curve. Curvature is always nonnegative; it is zero for straight-line motion and positive if the path is curving. This part of the acceleration is always perpendicular to the motion, and it corresponds to the acceleration one achieves in a car using the steering wheel to change the path of motion without necessarily changing the speed.

Example 2.16. Decompose the acceleration of the planar vector function $\boldsymbol{x}(t) = \langle t, t^2\rangle$ into its tangential and normal components. Also compute the unit tangent and unit normal vectors, the speed, and the curvature.

3 Mathematicians often object to using both $\boldsymbol{T}(t)$ and $\boldsymbol{T}(s)$ since then $\boldsymbol{T}(1)$ becomes ambiguous. (Is this $t = 1$ or $s = 1$?) This indeed can be a serious problem, but the problem will be avoided here by using a prime rather than a dot to indicate differentiation with respect to s and only writing $\boldsymbol{T}'$ as a function of s.

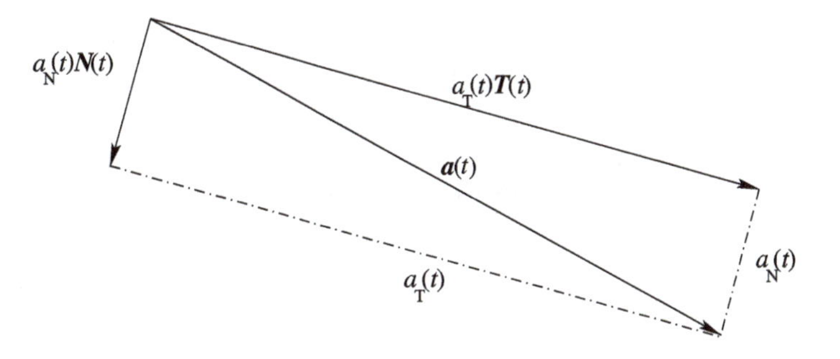

Figure 2.15: Acceleration and its normal and tangential components form a right triangle, thus the lengths of these vectors satisfy $|\boldsymbol{a}(t)|^2 = (a_T(t))^2 + (a_N(t))^2$.

Answer. By direct computation, $\boldsymbol{v}(t) = \langle 1, 2t\rangle$, $\boldsymbol{a}(t) = \langle 0, 2\rangle$, and the speed is $\dot{s}(t) = |v(t)| = \sqrt{1+4t^2}$. The unit tangent vector is

$$\boldsymbol{T}(t) = \frac{\boldsymbol{v}(t)}{|\boldsymbol{v}(t)|} = \frac{\langle 1, 2t\rangle}{\sqrt{1+4t^2}} = \left\langle \frac{1}{\sqrt{1+4t^2}}, \frac{2t}{\sqrt{1+4t^2}} \right\rangle.$$

The (scalar) tangential component of acceleration is

$$a_T(t) = \ddot{s}(t) = \frac{4t}{\sqrt{1+4t^2}}.$$

The curvature and the (scalar) normal component of acceleration can be computed directly, but it is more convenient to compute these indirectly from the results that have already been obtained. Since $\boldsymbol{a}(t)$ and its normal and tangential components form the sides of a right triangle (*cf.* Figure 2.15), the main tool in this indirect computation is the Pythagorean theorem. So

$$a_N(t) = \sqrt{|\boldsymbol{a}(t)|^2 - (a_T(t))^2} = \sqrt{4 - \frac{16t^2}{1+4t^2}} = \frac{2}{\sqrt{1+4t^2}}.$$

Since $a_N(t) = (\dot{s}(t))^2 \kappa(s(t))$ and $(\dot{s}(t))^2 = 1 + 4t^2$,

$$\kappa(s(t)) = \frac{2}{(1+4t^2)^{3/2}}.$$

Finally,

$$\boldsymbol{N}(t) = \boldsymbol{n}(s(t)) = \frac{\boldsymbol{a}(t) - a_T(t)\boldsymbol{T}(t)}{a_N(t)} = \frac{(0,2) - \frac{(4t,8t^2)}{1+4t^2}}{\frac{2}{\sqrt{1+4t^2}}} = \frac{\langle -2t, 1\rangle}{\sqrt{1+4t^2}}.$$

Notice how $\boldsymbol{n}(s(t))$ and $\boldsymbol{T}(t)$ are related—they are perpendicular as expected.

Example 2.17. For the circle centered at the origin with radius r (satisfying $x^2+y^2 = r^2$), please find the curvature $\kappa(s)$.

Answer. Because curvature is a function of arc length, it will be the same regardless of how one parameterizes the circle. So one should choose the simplest parameterization for computing derivatives: let $x = r\cos t$ and $y = r\sin t$. Then $\boldsymbol{x}(t) = \langle r\cos t, r\sin t\rangle$, and $\dot{s}(t) = |\boldsymbol{v}(t)| = r$ (constant for all t). Because $\dot{s}$ is constant, $a_T = \ddot{s} = 0$, and thus $a_N(t) = |\boldsymbol{a}(t)| = r$. Finally, since $a_N(t) = (\dot{s}(t))^2\kappa(s(t))$, one finds that $\kappa(s) = 1/r$. So not surprisingly, the curvature of a circle is constant and inversely proportional to its radius.

Exercises 2

2.1. For each of the following vector functions, $\boldsymbol{v}$, please determine its domain, and decide whether or not the indicated value of t_o is in the domain. Then provided the limit exists, please compute $\lim_{t\to t_o}\boldsymbol{v}(t)$:
(a) $\boldsymbol{v}(t) = \langle 1-5t^2, 14\pi e^{-2t}\rangle$, $t_o = 3$
(b) $\boldsymbol{v}(t) = \left\langle \frac{1}{1-t^2}, \frac{\sin t}{t}, \sqrt{4-t^2}\right\rangle$, $t_o = 0$
(c) $\boldsymbol{v}(t) = \left\langle \frac{1-t}{1-t^2}, \frac{\sin \pi t}{1-t}, \sqrt{1-t^2}\right\rangle$, $t_o = 1$
(d) $\boldsymbol{v}(t) = \left\langle \frac{1-\sqrt{t}}{1-t^2}, \frac{3t}{e^{t-1}-e^{1-t}}\right\rangle$, $t_o = 1$
(e) $\boldsymbol{v}(t) = \left\langle \frac{1-\sqrt{t}}{1-t^2}, \frac{3t}{e^{t-1}-e^{1-t}}, 14-9t+t^2\right\rangle$, $t_o = 7$

Answer. (b) The point $t_o = 0$ is not in the domain because of the second component, but $\lim_{t\to 0}\boldsymbol{v}(t) = \langle 1, 1, 2\rangle$.

2.2. For each of the following expressions $\boldsymbol{w}(t)$, please find its derivative $\dot{\boldsymbol{w}}(t)$. When necessary, please mention any restrictions on the domains of either $\boldsymbol{w}$ or $\dot{\boldsymbol{w}}$:
(a) $\boldsymbol{w}(t) = \langle t^3, te^t\rangle$
(b) $\boldsymbol{w}(t) = \langle \sin t^2, t\cos t^2, \mathrm{Tan}^{-1}(t^2)\rangle$
(c) $\boldsymbol{w}(t) = \langle \frac{1-t^2}{1+t^2}, \frac{1-e^{-t^2}}{1+t^2}\rangle$
(d) $\boldsymbol{w}(t) = \langle \ln(1+4t^2), t\ln(1+4t^2)\rangle$
(e) $\boldsymbol{w}(t) = \langle 3t^2-4t+7, e^2, \cosh(t^2)\rangle$
(f) $\boldsymbol{w}(t) = \langle t^2, t|t|\rangle$

Answer. (b) $\dot{\boldsymbol{w}}(t) = \langle 2t\cos t^2, \cos t^2 - 2t^2\sin t^2, \frac{2t}{1+t^4}\rangle$.

2.3. For $\boldsymbol{v}(t) = \langle e^{-t}\cos 2t, e^{-t}\sin 2t\rangle$, please find $\dot{\boldsymbol{v}}(t)$. Can you see an orthogonality relationship between $\boldsymbol{v}(t)$ and $\dot{\boldsymbol{v}}(t)$?

2.4. For $\boldsymbol{v}(\tau) = \langle 6\tau^2-5, \tan\tau\rangle$, please find the integral of $\boldsymbol{v}$ over the interval $[0, t]$. What is domain D of the function defined by this integral?

Answer. $\langle 2t^3-5t, -\ln(\cos t)\rangle$; $D = (-\pi/2, \pi/2)$.

2.5. Consider the vector equation $\dot{\boldsymbol{v}}(t) + \boldsymbol{v}(t) = \langle 1, 0\rangle$. Verify that $\boldsymbol{v}(t) = \langle 2e^{-t}+1, 3e^{-t}\rangle$ satisfies this equation along with the condition that $\boldsymbol{v}(0) = \langle 3, 3\rangle$.

2.6. For each parameterization $\gamma(t)$, please find an equation for the corresponding curve γ in $\mathbb{R}^2$ in terms of just x and y,

(a) $\gamma(t) = (5t^2, \sqrt{5}t), t \in [-3, 5]$
(b) $\gamma(t) = (2\sin t, \cos t), t \in [0, \pi]$
(c) $\gamma(t) = \left(\frac{1}{\sqrt{1+t^2}}, \frac{t}{\sqrt{1+t^2}} \right), t \in (-\infty, \infty)$

Answer. (b) $x^2 + 4y^2 = 4$ (an ellipse).

2.7. Please parameterize each of the following curves in $\mathbb{R}^2$ using sine and cosine functions:
(a) $x^2 + y^2 = 4$
(b) $(x+1)^2 + (y-5)^2 = 9$
(c) $3x^2 + 2y^2 = 12$

Answer. (a) $x = 2\cos t, y = 2\sin t, t \in [0, 2\pi)$. There are many other parameterizations.

2.8. Describe the curve γ in $\mathbb{R}^3$ traced out by the parameterization $\gamma(t) = (1 - t^2, t(1-t^2), t)$.

2.9. For the paramerization $\boldsymbol{x}(t) = \langle \cos t, \sin t, t \rangle$ of a circular helix, please find the velocity, speed, acceleration, unit tangent and unit normal vectors, and the curvature. Notice that for the direct calculation of the unit normal vector, only the vector portion of the derivative of the unit tangent vector is needed, since the scalar portion is cancelled out when one computes the length. Also notice that these calculations are much simpler for this circular helix then for the elliptical one in Example 2.11.

Answer. $\boldsymbol{v}(t) = \langle -\sin t, \cos t, 1 \rangle$, $\dot{s}(t) = |\boldsymbol{v}(t)| = \sqrt{2}$, $\boldsymbol{a}(t) = \langle -\cos t, -\sin t, 0 \rangle$, $\boldsymbol{T}(t) = \langle -\sin t, \cos t, 1 \rangle / \sqrt{2}$, $\boldsymbol{N}(t) = \langle -\cos t, -\sin t, 0 \rangle$.

2.10. Please find the velocity, speed, acceleration, and unit tangent and unit normal vectors for $\boldsymbol{x}(t) = \left\langle \pi t^2, \int_2^t \sin \tau \, d\tau \right\rangle$. Notice that the direct calculation of the unit normal vector is rather messy; how can this calculation be simplified?

2.11. Show that when speed is constant, acceleration is perpendicular to velocity.

2.12. For a baseball (particle) whose acceleration vector is $\boldsymbol{a}(t) = \langle 0, 0, -32 \rangle$, please find the velocity and position (give the most general answer possible). This is projectile motion neglecting air resistance and using $g = 32\,\text{ft/s}^2$ as the acceleration due to gravity. Notice that without air resistance, throwing a curve ball is not possible. **Hint:** Suppose that the initial position of the baseball is $\boldsymbol{x}(0) = \boldsymbol{x}_o = \langle a_o, b_o, c_o \rangle$, and the initial velocity is $\boldsymbol{v}(0) = \boldsymbol{v}_o = \langle u_o, v_o, w_o \rangle$ where all of the components of $\boldsymbol{x}_o$ and $\boldsymbol{v}_o$ are constant.

2.13.
(a) Please calculate velocity, speed, and the unit tangent vector $\boldsymbol{T}(t)$ for $\boldsymbol{x}(t) = \langle t^2, t^3 \rangle$.
(b) Now calculate the unit normal vector $\boldsymbol{N}(t)$, the tangential and normal components of acceleration, $\boldsymbol{a}_T(t)$ and $\boldsymbol{a}_N(t)$, and the curvature $\kappa(s(t))$.

Answer. (a) $\dot{s}(t) = |t|\sqrt{4+9t^2}$; (b) $\boldsymbol{a}_N(t) = 6t\langle -3t, 2\rangle/(4+9t^2)$, $\kappa(s(t)) = 6/t(4+9t^2)^{3/2}$.

2.14. Calculate velocity, speed, acceleration, $\boldsymbol{T}(t)$, $\boldsymbol{N}(t)$, the tangential and normal components of acceleration, and the curvature for the vector function $\boldsymbol{x}(t) = \langle 1 - t^2, t(1-t^2), t\rangle$ for $t \in [-2,2]$. **Hint:** Use the Pythagorean theorem to find $a_N(t)$. On the whole, this exercise is a rather messy calculation.

2.15. Please draw (either by hand or using your favorite calculator or computer software) the curves in Example 2.11, Example 2.12 and Example 2.14.

2.16. Suppose that a particle is moving along a curve so that its arc length is a linear function of time: $s(t) = \sigma t$ for some constant σ.
(a) Calculate the curvature for this curve in terms of acceleration.
(b) What is the direction of the acceleration relative to the curve?

Answer. (a) $\kappa(s) = |\boldsymbol{a}(s)|/\sigma^2$.

2.17. For $\boldsymbol{x}(t) = \langle t^2, 2t^3/3\rangle$, please find α, ω and the arc length of the curve traced out by this vector function $\boldsymbol{x}$, first for (a) $t \in [1,4]$, then for (b) $t \in [-2,2]$.

2.18. Consider the graph of $y = f(x) = x^2/2$ between the points $(0,0)$ and $(1,1/2)$. What is the length of this graph (its arc length as a curve)? **Hint:** One can always parameterize the graph of $y = f(x)$ as $y(t) = (t, f(t))$, that is, $x = t$ and $y = f(t)$, then choosing the appropriate values for the beginning and ending points in t.

2.19. Consider the curve traced out by the vector function $\boldsymbol{x}(t) = \langle t, t^2\rangle$, as in Example 2.16. Please find $s(t)$. **Hint:** Computing $s(t)$ by hand requires a trig substitution, writing $\sec^3\theta$ as $\sec\theta(1+\tan^2\theta)$, then a double integration by parts. To avoid all this, one can use the internet, computer software, or a good table of integrals.

2.20. Consider the helix traced out by the vector function $\boldsymbol{x}(t) = \langle 2\cos t, 2\sin t, 3t\rangle$. Find $s(t)$, $\boldsymbol{T}(t)$, $\boldsymbol{N}(t)$, $a_T(t)$, $a_N(t)$, and $\kappa(s)$.

2.21. Suppose that for a certain curve, $s(t) = 5t$ and curvature is constant $\kappa = 2$. Please find $|\boldsymbol{a}(t)|$ for the vector function that traces out this curve.

2.22. Why is $\boldsymbol{N}(t) = \boldsymbol{n}(s(t))$? **Hint:** Think about Equation 2.1.

2.23. Suppose that for $t > 0$, the arc length of a certain curve is $s(t) = t^2$. Please find the curvature for this curve as a function of arc length in terms of the length of acceleration.

Answer. $\kappa(s) = \sqrt{|\boldsymbol{a}(\sqrt{s})|^2 - 4}/4s$.

2.24. If $\boldsymbol{x}(t)$, $\dot{\boldsymbol{x}}(t)$ and $\ddot{\boldsymbol{x}}(t)$ all exist, please show that $\frac{d}{dt}(\boldsymbol{x}(t) \times \dot{\boldsymbol{x}}(t)) = \boldsymbol{x}(t) \times \ddot{\boldsymbol{x}}(t)$.

3 Multivariable derivatives—differentiation in $\mathbb{R}^n$

3.1 Limits in $\mathbb{R}^n$

The most important new concept introduced by calculus is the concept of a limit; "limit" is *the* thing that Leibniz and Newton never understood, and it was up to Cauchy to define the concept about a century later.[1] This section discusses limits in $\mathbb{R}^n$ and explores the similarities and differences between limits involving multivariable scalar functions and vector fields on the one hand, and those involving a single variable (including vector functions) on the other.

3.1.1 Definitions and the basics

Before we can define the multivariable limit, we need to state the definition of distance in $\mathbb{R}^n$:

Definition. The function $d : \mathbb{R}^n \times \mathbb{R}^n \to [0, \infty)$ such that $\forall\, \boldsymbol{x}, \boldsymbol{y} \in \mathbb{R}^n$,

$$d(\boldsymbol{x}, \boldsymbol{y}) := \sqrt{(x_1 - y_1)^2 + (x_2 - y_2)^2 + \cdots + (x_n - y_n)^2}$$

is the *distance* between $\boldsymbol{x}$ and $\boldsymbol{y}$.

Thus distance is a function that assigns a nonnegative real number to any two points or vectors $\boldsymbol{x}$ and $\boldsymbol{y}$ in $\mathbb{R}^n$, and when $n = 2$ or $n = 3$, this nonnegative real number is the classical Euclidean distance. Hence this is the traditional idea of distance, given in mathematical terms. For $n = 2$, the diagram in Figure 3.1 may be helpful, as should the next example.

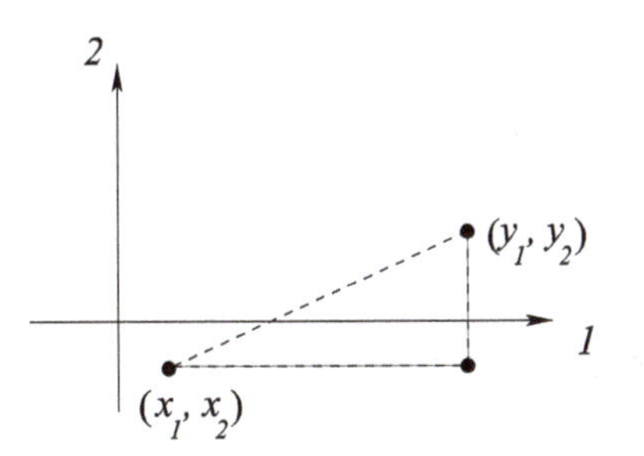

Figure 3.1: Distance in $\mathbb{R}^2$ where the dimensions and the axes are numbered. What point is at the corner of the right angle of the triangle? Where are $\boldsymbol{x}$ and $\boldsymbol{y}$? Where should $d(\boldsymbol{x}, \boldsymbol{y})$ be placed on this diagram?.

1 Augustin-Louis Cauchy (1789–1857) was a French mathematician, scientist, and engineer who worked to make the basic concepts of calculus understood via rigorous proof.

https://doi.org/10.1515/9783110660609-003

Example 3.1. For which values of β is the distance between the points $\boldsymbol{x} = (3,-1,2)$ and $\boldsymbol{y} = (\beta, 1-\beta, 3)$ less than 2? Which vector $\boldsymbol{v}$ has its tail at $\boldsymbol{x}$ and its head at $\boldsymbol{y}$, and what is its length?

Answer. We must decide for which β is $d(\boldsymbol{x},\boldsymbol{y}) < 2$, which means that $(3-\beta)^2 + (-1-(1-\beta))^2 + (2-3)^2 < 4$. Simplifying this inequality, one finds that it is equivalent to $\beta^2 - 5\beta + 5 < 0$. Since the roots of this quadratic are $(5 \pm \sqrt{5})/2$, we are looking for

$$\beta \in \left(\frac{5-\sqrt{5}}{2}, \frac{5+\sqrt{5}}{2}\right).$$

The vector with $(\beta, 1-\beta, 3)$ at its head and $(3,-1,2)$ at its tail is $\boldsymbol{v} = \langle \beta-3, 1-\beta-(-1), 3-2\rangle = \langle \beta - 3, 2-\beta, 1\rangle$. Finally, the length of this vector is exactly the distance between these two points: $|\boldsymbol{v}| = d(\boldsymbol{x},\boldsymbol{y}) = \sqrt{(3-\beta)^2 + (-1-(1-\beta))^2 + (2-3)^2} = \sqrt{2\beta^2 - 10\beta + 14}.$

An observant reader might have noticed that there is a connection between distance as it is defined here and the length of a difference vector (see Figure 3.2).

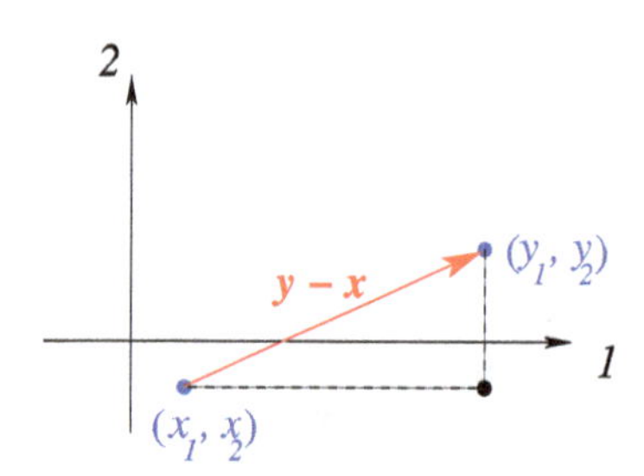

Figure 3.2: The length of $\boldsymbol{y} - \boldsymbol{x}$ is the same as $d(\boldsymbol{x}, \boldsymbol{y})$.

Proposition 3. *From the definitions above for distance and vector length,*

$$d(\boldsymbol{x},\boldsymbol{y}) = d(\boldsymbol{y},\boldsymbol{x}) = |\boldsymbol{x}-\boldsymbol{y}| = |\boldsymbol{y}-\boldsymbol{x}|.$$

Proof. From their definitions, all four of the expression in this proposition are equal to

$$\sqrt{(x_1-y_1)^2 + (x_2-y_2)^2 + \cdots + (x_n-y_n)^2} = \sqrt{(y_1-x_1)^2 + (y_2-x_2)^2 + \cdots + (y_n-x_n)^2}. \qquad \square$$

Our distance function can now be used to give a mathematical definition of limit. In everyday language, this definition precisely says that if $\boldsymbol{x}$ is sufficiently close to $\boldsymbol{x}_o$, then $\boldsymbol{f}(\boldsymbol{x})$ is as close as we want to a limiting value, $\boldsymbol{L}$.

Definition. Let $\boldsymbol{x}_0 \in \mathbb{R}^n$, and let $\boldsymbol{f} : \mathbb{R}^n \to \mathbb{R}^m$. Then $\boldsymbol{L} \in \mathbb{R}^m$ is the *limit* of $\boldsymbol{f}$ as $\boldsymbol{x}$ approaches $\boldsymbol{x}_0$ if and only if, given any $\epsilon > 0$, $\exists\, \delta > 0$ such that

$$0 < d(\boldsymbol{x},\boldsymbol{x}_0) < \delta \quad \Longrightarrow \quad d(\boldsymbol{f}(\boldsymbol{x}),\boldsymbol{L}) < \epsilon \,.$$

That is, if $0 < d(\boldsymbol{x},\boldsymbol{x}_0) < \delta$, then $d(\boldsymbol{f}(\boldsymbol{x}),\boldsymbol{L}) < \epsilon$.

Remarks.

1. If the reader has seen the ϵ, δ-definition of the limit in single-variable calculus, there is an important observation about this definition: *It is exactly the same as the single-variable case!* If the reader has only seen a more-intuitive development of the concept of limit, the important thing to understand is that this definition is just a mathematically-exact statement of the intuition. It says that whenever $\boldsymbol{x}$ is close to $\boldsymbol{x}_0$, then $\boldsymbol{f}(\boldsymbol{x})$ is close to $\boldsymbol{L}$. Or in other words, you tell me how close you want $\boldsymbol{f}(\boldsymbol{x})$ to be to $\boldsymbol{L}$ (the ϵ), and then I will tell you how close $\boldsymbol{x}$ must be to $\boldsymbol{x}_0$ to guarantee the result you want (the δ). A single-variable limit is depicted in Figure 3.3; a multivariable depiction is given in Figure 3.4.

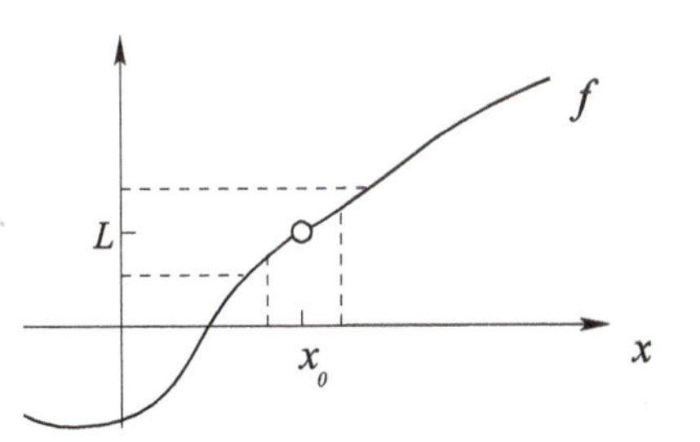

Figure 3.3: A single-variable limit. How close must x be to x_0 to guarantee that $f(x)$ is as close to L as we want—where do the δ and ϵ go in this diagram? Notice that the portion of the curve $y = f(x)$ above the interval on the x-axis lies entirely within the band centered on L coming across the y-axis.

2. As written, this definition only applies to functions defined for all $\boldsymbol{x} \in \mathbb{R}^n$; often $D := \text{Domain}(\boldsymbol{f}) \subsetneq \mathbb{R}^n$, that is, $\boldsymbol{f}$ is not defined at some points in $\mathbb{R}^n$. In this case, the definition is essentially the same, at least if D is a open set, except that $\boldsymbol{x}_0$ must be in or adjacent to D and $\boldsymbol{f} : D \to \mathbb{R}^m$.
3. $\boldsymbol{f}(\boldsymbol{x}_0)$ **is not necessarily defined!** It *might* be, but it does not have to be. Also $\boldsymbol{x}$ must be different from $\boldsymbol{x}_0$; the two must be close, but different.
4. For a given ϵ, if the limit exists, there will be *many* choices for δ. Indeed if a certain δ works, then so does $\delta/2$, $\delta/4$, $\delta/10$, and so forth.
5. **There is no multivariable l'Hôpital's rule!** If one wishes to use l'Hôpital's rule to study a multivariable limit, one must first reduce the problem to a single-variable limit.

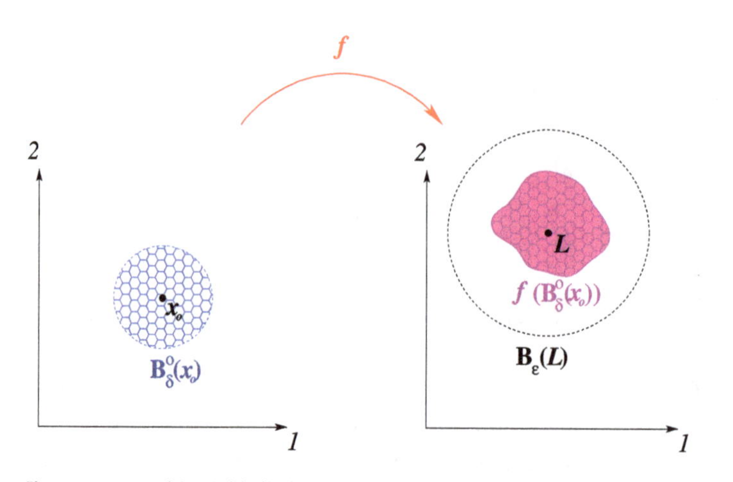

Figure 3.4: A multivariable limit. Here, the function $\boldsymbol{f}$ sends $\boldsymbol{x}$-values from a domain in $\mathbb{R}^2$ on the left to $\boldsymbol{y}$-values in some range in $\mathbb{R}^2$ on the right, that is, $\boldsymbol{y} = \boldsymbol{f}(\boldsymbol{x})$. Let $\boldsymbol{x}_o$ be any point in the domain on the left. The ball of radius δ centered at $\boldsymbol{x}_o$ excluding the center $\boldsymbol{x}_o$ itself is denoted as $B^\circ_\delta(\boldsymbol{x}_o)$. If the limit of $\boldsymbol{f}$ as $\boldsymbol{x} \to \boldsymbol{x}_o$ exists, then for some δ sufficiently small, $\boldsymbol{f}$ sends every point from $B^\circ_\delta(\boldsymbol{x}_o)$ into the ball of radius ϵ centred at $\boldsymbol{L}$, $B_\epsilon(\boldsymbol{L})$. In symbols, this is written $\boldsymbol{f}(B^\circ_\delta(\boldsymbol{x}_o)) \subset B_\epsilon(\boldsymbol{L})$. Again, where do the δ and ϵ go in this diagram?.

Notation. If the limit of $\boldsymbol{f}$ as $\boldsymbol{x}$ approaches $\boldsymbol{x}_o$ exists and equals $\boldsymbol{L}$, then we can write

$$\lim_{\boldsymbol{x}\to\boldsymbol{x}_o} \boldsymbol{f}(\boldsymbol{x}) = \boldsymbol{L}$$

just as in the single-variable case.

Example 3.2. Some limits are rather easy to evaluate without carefully referring to the definition. For example, if $\boldsymbol{x} = (x, y) \in \mathbb{R}^2$, to evaluate

$$\lim_{(x,y)\to(1,-\pi)} x\cos(y) + y$$

one needs only to note that as y approaches $-\pi$, $\cos(y)$ approaches -1. Since x is approaching 1, the entire expression is approaching $1(-1)-\pi = -1-\pi$, and thus $L = -1-\pi$. Notice that in evaluating this limit we have implicitly used that cosine and indeed the entire expression is continuous. Continuity will be discussed in the next section.

One can use our δ-ϵ definition to know that a limit exists for a specific function, but this is not the most important reason for having a clear, mathematical definition. The real advantage of such a definition is that it can be used to prove that limits have certain important properties, for example, we have the following:

Theorem 1. *If* $\lim_{\boldsymbol{x}\to\boldsymbol{x}_o} \boldsymbol{f}(\boldsymbol{x})$ *exists, then its value L is unique.*

Proof. Suppose there are two values, that is, suppose that $\lim_{\boldsymbol{x}\to\boldsymbol{x}_o} \boldsymbol{f}(\boldsymbol{x}) = \boldsymbol{L}_1$ and also $\lim_{\boldsymbol{x}\to\boldsymbol{x}_o} \boldsymbol{f}(\boldsymbol{x}) = \boldsymbol{L}_2$. From the definition of limit, given any $\epsilon > 0$, $\exists\, \delta > 0$ such that when $0 < d(\boldsymbol{x}, \boldsymbol{x}_0) < \delta$, then $d(\boldsymbol{f}(\boldsymbol{x}), \boldsymbol{L}_1) < \epsilon/2$ and $d(\boldsymbol{f}(\boldsymbol{x}), \boldsymbol{L}_2) < \epsilon/2$. But from the triangle inequality,

$$\begin{aligned} d(\boldsymbol{L}_1, \boldsymbol{L}_2) &\leq d(\boldsymbol{L}_1, \boldsymbol{f}(\boldsymbol{x})) + d(\boldsymbol{f}(\boldsymbol{x}), \boldsymbol{L}_2) \\ &< \quad \epsilon/2 \quad + \quad \epsilon/2 \\ &= \epsilon \end{aligned}$$

So $d(\boldsymbol{L}_1, \boldsymbol{L}_2)$ is less than any positive real number, which means that $d(\boldsymbol{L}_1, \boldsymbol{L}_2) = 0$. This is only possible if $\boldsymbol{L}_1 = \boldsymbol{L}_2$, so there can not be two distinct values for the limit. □

3.1.2 0/0 indeterminate form

Again, as in the single-variable case, one of the most interesting situations occurs when there is a quotient with the numerator and denominator both going to zero as $\boldsymbol{x}$ approaches $\boldsymbol{x}_0$. This situation is called a 0/0 *indeterminate form*. Other indeterminate forms are possible in the multivariable case, but since derivatives are always of the 0/0 indeterminate form, this is the only indeterminate form studied here. We will first consider limits of this form when there is no single limiting value, implying that the limit does not exist. Later we will discuss computing limits of this form when a single limiting value does exist.

When the limit does not exist (DNE)

The uniqueness of the limit value as proven in Theorem 1 has a very important practical consequence: if one obtains distinct values for a limit as one approaches $\boldsymbol{x}_0 \in \mathbb{R}^n$ along distinct paths, then the limit in fact does not exist. This sort of thing happens with single-variable limits, too—when the value from the left differs from the value from the right. But in the multivariable case, there are an uncountable number of ways to approach $\boldsymbol{x}_0$ (see Figure 3.5), and for the limit to exist, *all* of these approach paths must yield the same limiting value.

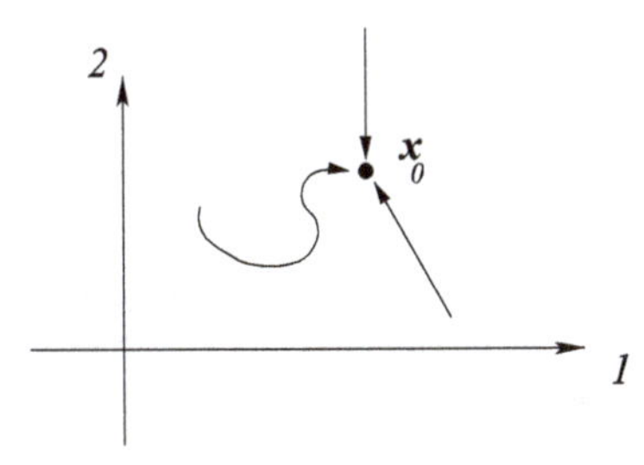

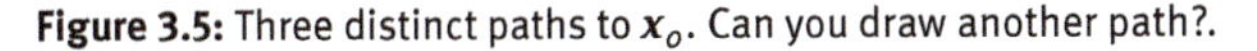

Figure 3.5: Three distinct paths to $\boldsymbol{x}_0$. Can you draw another path?.

Example 3.3 (Railroad underpass). Consider

$$\lim_{(x,y)\to(0,0)} \frac{xy}{x^2+y^2}$$

In this case, both the numerator and the denominator go to zero as $(x,y) \to (0,0)$. So this limit is in 0/0 indeterminate form.

Consider the value as one approaches the origin along the path $y = 0, x > 0, x \to 0$. Along this path, the numerator is always zero, while the denominator is positive. So

$$\lim_{x\to 0^+,\, y=0} \frac{xy}{x^2+y^2} = \lim_{x\to 0^+} \frac{x(0)}{x^2+0^2} = \lim_{x\to 0^+} \frac{0}{x^2} = \lim_{x\to 0^+} 0 = 0$$

Now try the path $x = 0, y > 0, y \to 0$; the same sort of analysis leads one to

$$\lim_{\substack{x\to 0^+\\ y=0}} \frac{xy}{x^2+y^2} = \lim_{x\to 0^+} \frac{0}{x^2} = 0\,.$$

So at first glance, one *might* be lead to suspect that this limit exists and $L = 0$. But two paths are surely not all paths. Consider the path $y = x, x > 0, x \to 0$. Here,[2]

$$\lim_{\substack{x\to 0^+\\ y=x}} \frac{xy}{x^2+y^2} = \lim_{x\to 0^+} \frac{x(x)}{x^2+x^2} = \lim_{x\to 0^+} \frac{x^2}{2x^2} = \lim_{x\to 0^+} \frac{1}{2} = 1/2$$

Thus since the value of this limit would depend on the path of approach, the limit itself does not exist (DNE).

Where does the name of this problem come from? A three-dimensional graph (see Figure 3.6) of the expression in this problem can answer this question, as well as help make clear why this limit is path-dependent. Imagine two railroad lines traveling on the three-dimensional surface $z = xy/(x^2+y^2)$, one on $y = x$, and the other on $y = -x$. Notice that a (two-dimensional, zero-width) train on the $y = -x$ line will pass under another train on the $y = x$ line as both cross through $(0,0)$.

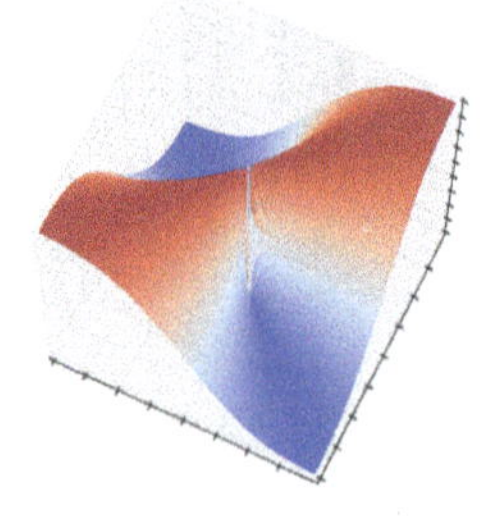

Figure 3.6: The surface $z = xy/(x^2+y^2)$ near the origin. On the deep red ridge, $f(x,x) = 1/2$ for $x \neq 0$; in the deep blue troff, $f(x,-x) = -1/2$ again for $x \neq 0$. At $x = 0$, the surface is undefined, and there is a infinitesimal slit in the surface.

2 Keep in mind that one can divide through by x^2 exactly because $x \neq 0$.

There are, of course, other ways that limits can fail to exist, for example, if $f(\boldsymbol{x})$ becomes unbounded or oscillates wildly as $\boldsymbol{x}$ approaches $\boldsymbol{x}_0$. These cases may be familiar from single-variable calculus, but they will also be discussed in the section on types of discontinuities below.

When the limit does exist

The above example more or less gives the method for showing that a limit does not exist: find two or more paths of approach along which the limit would either not exist or take on differing values. Unfortunately, it is often more difficult to show that a limit that exists really does exist. To begin with, it is never possible to show that a limit does exist by considering two or more paths–it always might be the case that there is still another path where the limit would approach a distinct value. One can never check *every* possible path separately because there are uncountably many of them.

The next two examples present two key principles for evaluating limits that *do* exist. The first is to reduce (if possible) a multivariable limit to a single-variable limit. The second is to use polar coordinates (or spherical coordinates in three or more dimensions). Neither of these techniques works in all cases, but both are useful for a wide variety of problems.

Example 3.4. Consider

$$\lim_{\substack{(x,y)\to(0,0)\\ x\neq 0}} \frac{\sin xy}{x}.$$

Again, both the numerator and the denominator go to zero as $(x,y) \to (0,0)$, so this limit is also in indeterminate form. Since the numerator is a trig expression while the denominator is a (very simple) polynomial, there is no way to simplify this quotient. On the other hand, the quotient is reminiscent of $\sin x/x$ and the limit of this latter quotient is a famous one:

$$\lim_{x\to 0} \frac{\sin x}{x} = 1$$

(see almost any single-variable calculus text for a proof of this result). Can knowledge of this latter limit be used to compute the limit in this example? The answer is "yes," and to see how to do this, one must recall that the limit of a product is the product of the limits. This allows regrouping after multiplying the numerator and the denominator by y provided that $y \neq 0$, followed by the substitution $u = xy$:

$$\begin{aligned}
\lim_{\substack{(x,y)\to(0,0)\\ x\neq 0}} \frac{\sin xy}{x} &= \left(\lim_{\substack{(x,y)\to(0,0)\\ x,y\neq 0}} \frac{\sin xy}{xy}\right)\left(\lim_{(x,y)\to(0,0)} y\right)\\
&= \left(\lim_{u\to 0} \frac{\sin u}{u}\right)\left(\lim_{(x,y)\to(0,0)} y\right)\\
&= \qquad (1) \qquad\qquad (0) \qquad = 0
\end{aligned}$$

The substitution $u = xy$ allows one to treat the first limit in the product as a single-variable limit, making it relatively easy to compute. Notice that writing the limit of a product as the product of the limits depends on the latter two limits both existing, which fortunately they do in this example. Also one should note that $y = 0$ *is* allowed in the original limit, so this case must be considered, too. Again fortunately $\sin xy/x$ is zero when $y = 0$.

The computation in the previous example required the following technical proposition.

Proposition 4. *Let $\boldsymbol{f}, \boldsymbol{g} : D \subset \mathbb{R}^n \to \mathbb{R}^m$ for some open domain D, let α be a scalar valued multivariable function: $\alpha : D \to \mathbb{R}$, and let $\mathbf{x}_o$ either be in or adjacent to D. Provided each of the limits on the right exist and there is no division by zero, each of the following equalities hold:*

- $\lim_{\mathbf{x}\to\mathbf{x}_o} (\boldsymbol{f}(\mathbf{x}) \pm \mathbf{g}(\mathbf{x})) = \lim_{\mathbf{x}\to\mathbf{x}_o} \boldsymbol{f}(\mathbf{x}) \pm \lim_{\mathbf{x}\to\mathbf{x}_o} \mathbf{g}(\mathbf{x})$
- $\lim_{\mathbf{x}\to\mathbf{x}_o} \alpha(\mathbf{x})\boldsymbol{f}(\mathbf{x}) = (\lim_{\mathbf{x}\to\mathbf{x}_o} \alpha(\mathbf{x}))(\lim_{\mathbf{x}\to\mathbf{x}_o} \boldsymbol{f}(\mathbf{x}))$
- $\lim_{\mathbf{x}\to\mathbf{x}_o} (\boldsymbol{f}(\mathbf{x}) * \mathbf{g}(\mathbf{x})) = \lim_{\mathbf{x}\to\mathbf{x}_o} \boldsymbol{f}(\mathbf{x}) * \lim_{\mathbf{x}\to\mathbf{x}_o} \mathbf{g}(\mathbf{x})$
  Notice that unless $m = 1$, this product $(*)$ must be understood as a dot product, cross product or some other vector product.
- *For $m = 1$, $\lim_{\mathbf{x}\to\mathbf{x}_o} \left(\frac{f(\mathbf{x})}{g(\mathbf{x})}\right) = \frac{\lim_{\mathbf{x}\to\mathbf{x}_o} f(\mathbf{x})}{\lim_{\mathbf{x}\to\mathbf{x}_o} g(\mathbf{x})}$ provided that $\lim_{\mathbf{x}\to\mathbf{x}_o} g(\mathbf{x}) \neq 0$*

Proof. Let us start with the limit of a sum: Since the limits on the right exist, let $\boldsymbol{L}_f := \lim_{\mathbf{x}\to\mathbf{x}_o} \boldsymbol{f}(\mathbf{x})$ and $\boldsymbol{L}_g := \lim_{\mathbf{x}\to\mathbf{x}_o} \mathbf{g}(\mathbf{x})$. Given any $\epsilon > 0$, pick δ sufficiently small so that when $d(\mathbf{x}, \mathbf{x}_o) < \delta$, both $d(\boldsymbol{f}(\mathbf{x}), \boldsymbol{L}_f) < \epsilon/2$ and $d(\mathbf{g}(\mathbf{x}), \boldsymbol{L}_g) < \epsilon/2$ (this is what the definition of limit gives us). Then by Proposition 3 and the triangle inequality (Proposition 1),

$$\begin{aligned} d(\boldsymbol{f}(\mathbf{x}) + \mathbf{g}(\mathbf{x}), \boldsymbol{L}_f + \boldsymbol{L}_g) &= |(\boldsymbol{f}(\mathbf{x}) + \mathbf{g}(\mathbf{x}) - (\boldsymbol{L}_f + \boldsymbol{L}_g)| = |(\boldsymbol{f}(\mathbf{x}) - \boldsymbol{L}_f) - (\boldsymbol{L}_g - \mathbf{g}(\mathbf{x}))| \\ &\leq |\boldsymbol{f}(\mathbf{x}) - \boldsymbol{L}_f| + |\boldsymbol{L}_g - \mathbf{g}(\mathbf{x})| = d(\boldsymbol{f}(\mathbf{x}), \boldsymbol{L}_f) + d(\mathbf{g}(\mathbf{x}), \boldsymbol{L}_g) < \epsilon/2 + \epsilon/2 = \epsilon . \end{aligned}$$

So, again by the definition of limit, $\lim_{\mathbf{x}\to\mathbf{x}_o} (\boldsymbol{f}(\mathbf{x}) + \mathbf{g}(\mathbf{x})) = \boldsymbol{L}_f + \boldsymbol{L}_g = \lim_{\mathbf{x}\to\mathbf{x}_o} \boldsymbol{f}(\mathbf{x}) + \lim_{\mathbf{x}\to\mathbf{x}_o} \mathbf{g}(\mathbf{x})$.

The proofs of all the other parts of this proposition are left as exercises (see Exercise 3.7). □

Example 3.5. Please compute the following vector field limit:

$$\lim_{\substack{(x,y)\to(0,0) \\ x,y>0}} \left\langle \frac{\sqrt{3+xy} - \sqrt{3-xy}}{3xy}, \frac{e^{3xy}}{\sqrt{3+xy} + \sqrt{3-xy}} \right\rangle$$

where the limit is restricted to positive values of x and y.

Answer. Notice that since limits of vector expressions are computed componentwise, there are in fact two separate limits to consider here. The limit for the first component is of 0/0 indeterminate form; for the second component, there actually is no indeterminate form. One could proceed to compute these limits separately, but the expression in the second component gives a hint as to what to do with the first component:

$$\begin{aligned}
&\left\langle \frac{\sqrt{3+xy}-\sqrt{3-xy}}{3xy}, \frac{e^{3xy}}{\sqrt{3+xy}+\sqrt{3-xy}} \right\rangle \\
&= \frac{1}{\sqrt{3+xy}+\sqrt{3-xy}} \left\langle \frac{(\not{3}+xy)-(\not{3}-xy)}{3xy}, e^{3xy} \right\rangle \\
&= \frac{1}{\sqrt{3+xy}+\sqrt{3-xy}} \left\langle \frac{2}{3}, e^{3xy} \right\rangle
\end{aligned}$$

provided that no denominator is zero. The key is that the numerator of the first component and the denominator of the second are conjugates, so multiplying the numerator and denominator of the first component by the sum of the square roots allows us to simply the first component. So using the second point from Proposition 4 above, we can compute

$$\begin{aligned}
&\lim_{\substack{(x,y)\to(0,0)\\ x,y>0}} \left\langle \frac{\sqrt{3+xy}-\sqrt{3-xy}}{3xy}, \frac{e^{3xy}}{\sqrt{3+xy}+\sqrt{3-xy}} \right\rangle \\
&= \left(\lim_{\substack{(x,y)\to(0,0)\\ x,y>0}} \frac{1}{\sqrt{3+xy}+\sqrt{3-xy}} \right) \left(\lim_{\substack{(x,y)\to(0,0)\\ x,y>0}} \left\langle \frac{2}{3}, e^{3xy} \right\rangle \right) \\
&= \left(\sqrt{3}/18\right) \langle 2, 3 \rangle \ .
\end{aligned}$$

Example 3.6. Consider

$$\lim_{(x,y)\to(0,0)} \frac{x^2 y}{x^2+y^2}$$

Notice that this looks rather like the first example of this section, Example 3.3, where the limit did not exist. Is the story the same here? The answer is "no," and perhaps the best way to see this is to use polar coordinates. Let

$$x = r\cos\theta \quad y = r\sin\theta$$

Notice that $(x,y) \to (0,0) \iff r \to 0$. So

$$\begin{aligned}
\lim_{(x,y)\to(0,0)} \frac{x^2 y}{x^2+y^2} &= \lim_{r\to 0} \frac{(r\cos\theta)^2(r\sin\theta)}{(r\cos\theta)^2+(r\sin\theta)^2} \\
&= \lim_{r\to 0} r(\cos\theta)^2(\sin\theta)
\end{aligned}$$

As r goes to zero, the value of theta may vary between 0 and 2π, but both $\cos\theta$ and $\sin\theta$ are bounded between -1 and 1. Thus the extra power of r in the numerator forces the limit to zero:

$$\lim_{(x,y)\to(0,0)} \frac{x^2 y}{x^2+y^2} = 0$$

By the way, it is worth noting that polar coordinates could also have been used in the first example above where the limit DNE:

$$\lim_{(x,y)\to(0,0)} \frac{xy}{x^2+y^2} = \lim_{r\to 0} \cos\theta \sin\theta = \cos\theta\sin\theta$$

since the expression in the limit does not depend on r. But since the apparent value of the limit depends on θ (i. e., the direction of approach), this limit can not exist.

The above examples give some sense of how to compute limits that appear to exist, but they do not attempt to show how to handle all possible cases where a multivariable limit is in indeterminate form. Other techniques may be needed. For example, algebraic simplifications are often helpful in the multivariable case as it is in the single-variable case, and although there is no multivariable l'Hopital's rule, one can still use this tool after reducing a multivariable limit to a single-variable limit as in the second example above.

Example 3.7. Consider

$$\lim_{\substack{(x,y)\to(1,1)\\ x\neq y}} \frac{x^2-y^2}{x-y}$$

Putting this limit into standard polar coordinates does not help, and considering different paths does not help either, since this limit does exist. The key step in simplifying this limit is factoring the numerator and cancelling the common factor of $x-y$. After doing this, one finds that this limit is 2.

3.1.3 Something that does not work

A tempting approach for trying to show that a limit exists is to show that $f(\boldsymbol{x})$ approaches the same limiting value L as $\boldsymbol{x} \to \boldsymbol{x}_o$ along *every* line that passes through $\boldsymbol{x}_o$. Specifically, if $f : \mathbb{R}^2 \to \mathbb{R}$, and if $\boldsymbol{x}_o = (0,0)$, one would consider $y = mx$ for all values of m and show that

$$\lim_{(x,y)\to(0,0)} f(x,y) = \lim_{x\to 0} f(x,mx) = L$$

independent of m. This idea is appealing, but unfortunately it is not sufficient. The limit must take on the same value as $\boldsymbol{x} \to \boldsymbol{x}_o$ in any manner, along *any* path, not just along lines.

Consider the follow counterexample: Suppose $f(x,y) = 0$ for all $(x,y) \in \mathbb{R}^2$ except for $x^2/2 < y < 2x^2$. For the region bounded by these two parabolas, suppose that $f = 1$. Then the limit along[3] $y = mx$ is 0 for every $m > 0$, because every such line is above the

3 One would also have to consider the line $x = 0$ separately.

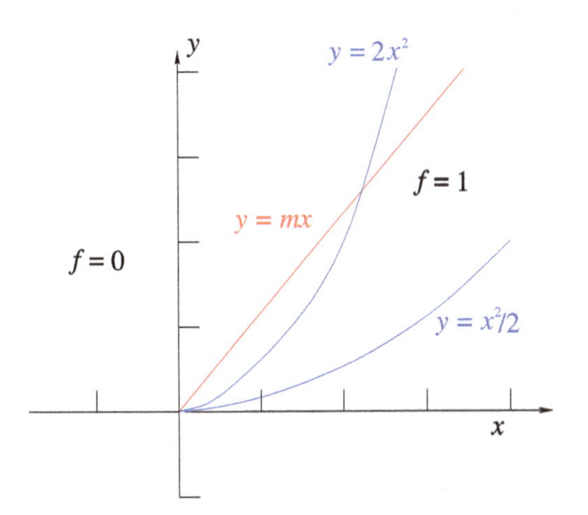

Figure 3.7: Why testing along lines is not enough to show that a limit exists. Because the line $y = mx$ (in red) is above the upper parabola for every $m > 0$ when x is sufficiently close to 0, the limit of f is 0 along any of these lines. But the limit is 1 along $y = x^2$, so in general, the limit DNE.

upper parabola for x near zero (see Figure 3.7). On the parabola $y = x^2$ when $x > 0$, however, $f(x, x^2) = 1$. So the limit,

$$\lim_{(x,y)\to(0,0)} f(x,y) \quad \text{DNE}\,.$$

3.2 Continuity in $\mathbb{R}^n$

3.2.1 Definition and examples

Now we turn our attention to the concept of continuity, a concept very related to that of limit. In fact, at first glance, the definitions look the same. Can you spot the two differences?

Definition. For some open domain $D \subset \mathbb{R}^n$, suppose that $\boldsymbol{f} : D \to \mathbb{R}^m$. Then $\boldsymbol{f}$ is *continuous* at $\boldsymbol{x}_o \in D$ if and only if given any $\epsilon > 0$, $\exists\, \delta > 0$ such that

$$d(\boldsymbol{x}, \boldsymbol{x}_o) < \delta \quad \Longrightarrow \quad d(\boldsymbol{f}(\boldsymbol{x}), \boldsymbol{f}(\boldsymbol{x}_o)) < \epsilon\,.$$

That is, if $d(\boldsymbol{x}, \boldsymbol{x}_o) < \delta$, then $d(\boldsymbol{f}(\boldsymbol{x}), \boldsymbol{f}(\boldsymbol{x}_o)) < \epsilon$.

Remarks.

1. This definition is just like the definition of the limit, except that here $d(\boldsymbol{x}, \boldsymbol{x}_o) = 0$ is possible, so we must now consider $\boldsymbol{x} = \boldsymbol{x}_o$, and thus $\boldsymbol{f}(\boldsymbol{x}_o)$ must be defined. In addition, now the value of the limit must be the value of the function: $\boldsymbol{L} = \boldsymbol{f}(\boldsymbol{x}_o)$. This is all exactly the same as the single-variable case (see Figure 3.8).
2. If $\lim_{\boldsymbol{x}\to\boldsymbol{x}_o} \boldsymbol{f}(\boldsymbol{x})$ DNE, then $\boldsymbol{f}$ cannot possibly be continuous at $\boldsymbol{x}_o$, and there is no need for any further consideration. So since the limit DNE in the first example

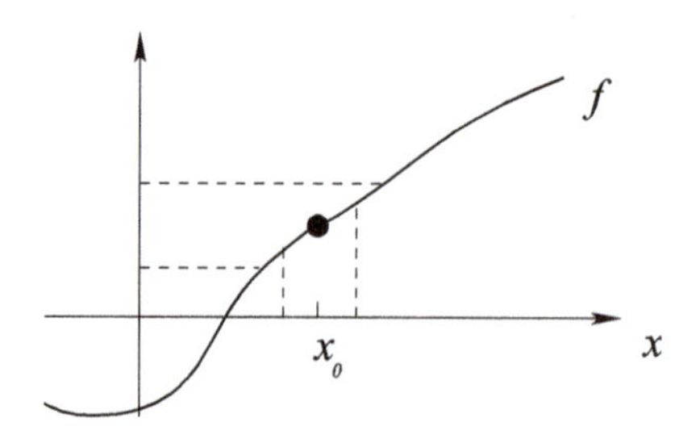

Figure 3.8: Single-variable continuity. This diagram tells the story in both the single-variable and multivariable cases. Where is $\boldsymbol{L}$ in this case? Where are δ and ϵ?.

in the previous section, there is no need to consider further whether or not the railroad underpass function is continuous at the origin. On the other hand, if $\lim_{\boldsymbol{x}\to\boldsymbol{x}_0} \boldsymbol{f}(\boldsymbol{x})$ exists and has the value $\boldsymbol{L}$, then $\boldsymbol{f}$ can be made continuous at $\boldsymbol{x}_0$ (if it is not already) by changing the value of $\boldsymbol{f}(\boldsymbol{x}_0)$ to $\boldsymbol{L}$.

3. Notice that $\boldsymbol{f}$ is continuous at $\boldsymbol{x}_0$ if and only if

$$\lim_{\boldsymbol{x}\to\boldsymbol{x}_0} \boldsymbol{f}(\boldsymbol{x}) = \boldsymbol{f}(\lim_{\boldsymbol{x}\to\boldsymbol{x}_0} \boldsymbol{x}) = \boldsymbol{f}(\boldsymbol{x}_0).$$

That is, continuity is equivalent to being able to exchange the order of taking the limit and evaluating the function.

Example 3.8. In the previous section, we found that

$$\lim_{(x,y)\to(1,-\pi)} x\cos(y) + y = -(1+\pi).$$

Since this is indeed the value that one finds by plugging $(1,-\pi)$ into $x\cos(y) + y$, the function $f(x,y) = x\cos(y) + y$ is continuous at $(1,-\pi)$. Unfortunately, many limit/continuity problems are not as easy as this.

Example 3.9. Now let us reconsider a previous expression:

$$\frac{\sin xy}{x}$$

In the previous section, we saw that

$$\lim_{\substack{(x,y)\to(0,0)\\ x\neq 0}} \frac{\sin xy}{x} = 0$$

but this is, in itself, not enough to make $\frac{\sin xy}{x}$ continuous. The problem is that as written, $\frac{\sin xy}{x}$ is undefined at the origin (do you see why?). Still since the limit exists, it is possible to overcome this problem by extending the domain to include the line $x = 0$. Thus if one defines

$$f(x,y) = \begin{cases} \frac{\sin xy}{x} & x \neq 0 \\ y & x = 0 \end{cases}$$

then this function f is continuous, even when $x = 0$.

The next proposition is really just a restatement of Proposition 4 where now the value of each limit is the value of the function at $\mathbf{x} = \mathbf{x}_o$.

Proposition 5. *For some open domain $D \subset \mathbb{R}^n$, suppose that $\mathbf{f}, \mathbf{g} : D \to \mathbb{R}^m$ are both continuous at $\mathbf{x}_o \in D$, and suppose that $\alpha : D \to \mathbb{R}$ is a scalar valued multivariable function that is also continuous at $\mathbf{x}_o$. Provided that there is no division by zero, then:*
- *$(\mathbf{f} \pm \mathbf{g})(\mathbf{x}) := \mathbf{f}(\mathbf{x}) \pm \mathbf{g}(\mathbf{x})$ is continuous at $\mathbf{x}_o$.*
- *$(\alpha \mathbf{f})(\mathbf{x}) := \alpha(\mathbf{x})\mathbf{f}(\mathbf{x})$ is continuous at $\mathbf{x}_o$.*
- *$(\mathbf{f} * \mathbf{g})(\mathbf{x}) := \mathbf{f}(\mathbf{x}) * \mathbf{g}(\mathbf{x})$ is continuous at $\mathbf{x}_o$.*
  Again, unless $m = 1$, the product $(*)$ must be understood as some sort of vector product, for example, a dot product.
- *For $m = 1$, $\left(\frac{f}{g}\right)(\mathbf{x}) = \frac{f(\mathbf{x})}{g(\mathbf{x})}$ is continuous at $\mathbf{x}_o$.*

Remark. Proposition 5 may seem formulaic, even trivial, but it has many important consequences. For example, it means that any polynomial or vector with polynomial components must be continuous no matter how many variables are involved.

Example 3.10. Suppose $f : D \subset \mathbb{R}^2 \to \mathbb{R}$ by

$$f(x,y) = \frac{xy^2 + x^2 - 4xy + 2x + 1}{x^2 + y^2 - 2x - 4y + 5}$$

with D being all of $\mathbb{R}^2$ where the denominator is not zero. Is this function f continuous on its domain, and can it be extended to be continuous for all of $\mathbb{R}^2$?

Answer. Since f is a rational function, by Proposition 5, f is continuous except at points where the denominator is zero, so f is continuous on its domain. For this rational function, since the denominator is quadratic with no cross terms (terms having both x and y), the zeros of the denominator can be determined by completing the square with respect to both x and y:

$$x^2 + y^2 - 2x - 4y + 5 = (x^2 - 2x + 1) + (y^2 - 4y + 4) = (x-1)^2 + (y-2)^2$$

Thus the only point not in the domain of f is $(x,y) = (1,2)$. If the numerator is nonzero at $(x,y) = (1,2)$, then there is no way to extend the definition of f at this point to make it continuous. For this f, however, the numerator is in fact zero at $(x,y) = (1,2)$, so

$$\lim_{(x,y)\to(1,2)} f(x,y)$$

is in $0/0$ indeterminate form.

To determine how f behaves near $(x,y) = (1,2)$, it is helpful to transform the domain variables; define

$$\tilde{x} := x - 1 \quad \tilde{y} := y - 2 .$$

Substituting $x = \tilde{x} + 1$ and $y = \tilde{y} + 2$ into f, one finds that

$$f(\tilde{x}+1, \tilde{y}+2) = \frac{\tilde{x}\tilde{y}^2 + \tilde{x}^2 + \tilde{y}^2}{\tilde{x}^2 + \tilde{y}^2} = \frac{\tilde{x}\tilde{y}^2}{\tilde{x}^2 + \tilde{y}^2} + 1,$$

and thus using polar coordinates $\tilde{x} = \tilde{r}\cos\tilde{\theta}$ and $\tilde{y} = \tilde{r}\sin\tilde{\theta}$

$$\lim_{(x,y)\to(1,2)} f(x,y) = \lim_{(\tilde{x},\tilde{y})\to(0,0)} f(\tilde{x}+1, \tilde{y}+2) = \lim_{\tilde{r}\to 0}(\tilde{r}\cos\tilde{\theta}\sin^2\tilde{\theta} + 1) = 1.$$

The value of this limit was obtained by introducing polar coordinates as in Example 3.6. Hence this function can be extended to be continuous for all of $\mathbb{R}^2$:

$$f(x,y) = \begin{cases} \frac{xy^2+x^2-4xy+2x+1}{x^2+y^2-2x-4y+5} & (x,y) \neq (1,2) \\ 1 & (x,y) = (1,2) \end{cases}$$

3.2.2 Types of discontinuities

For real-valued multivariable functions, there are four main types of discontinuities: removable, jump, pole, and essential. It is possible to combine these to achieve a discontinuities not of any one of these types, but nonetheless these four are important to know.

Examples of removable discontinuities are given in Examples 3.9 and 3.10. There the function or expression is defined and continuous except at a single point or along a curve. But in addition, limits approaching this point or curve exist and are all equal, and one can use the values of these limits to "remove" the discontinuity and extend the definition of the function or expression continuously.

Jump discontinuities, in contrast, cannot be removed. These exist typically along a curve, and while the one-sided limits exist from both sides up to this curve, the limiting values are different on the two sides. Such a function was given in the example in Section 3.1.3; there the function takes on the value 1 on one side of the parabolas and 0 on the other side.

In contrast to jumps, poles typically occur at points, and more importantly, the function or expression becomes unbounded as this point is approached, that is, the function or expression diverges to $\pm\infty$ as the point is approached. A simple example of a pole is

$$f(x,y) = \frac{1}{x^2 + y^2}$$

which is defined on all of $\mathbb{R}^2$ except at the origin. As the origin is approached, this f diverges to $+\infty$.

The final discontinuity type, essential discontinuities, is characterized by wild oscillations near a point or curve. Perhaps the best example of an essential discontinuity is the function

$$f(x,y) = \sin\left(\frac{1}{x^2+y^2}\right)$$

near $(x,y) = (0,0)$. This f has all of the oscillation of $\sin(r^2)$ for $r^2 \in (1,+\infty)$ compressed into the circular disk of radius 1 centred on the origin.

3.2.3 Piecewise continuity

There is one more concept regarding continuity that is needed: piecewise continuity.

Definition. For some finite, closed interval $I \subset \mathbb{R}$, a real-valued function $f : I \to \mathbb{R}$ is *piecewise continuous* on I iff f is continuous on I except at a finite number of points, and at these points, the only discontinuities are jump and removable discontinuities. Thus $f(x)$ has finite limits as x approaches any of the points of discontinuity.

3.3 The derivative in $\mathbb{R}^n$

Everyone who has taken a basic calculus course *should* be familiar with the following definition.

Definition. For some open domain $D \subset \mathbb{R}$, suppose that $f : D \to \mathbb{R}$. Then f is *differentiable* at $x_o \in D$ iff the following limit exists:

$$\lim_{x \to x_o} \frac{f(x) - f(x_o)}{x - x_o}.$$

When this limit exists, it is called the *derivative* and denoted by $f'(x_o) \equiv \frac{df}{dx}(x_o)$.

This definition for the derivative in $\mathbb{R}$ is central to everything in single-variable calculus, but unfortunately, it is not easy to extend directly to the multivariable case where $n > 1$. To reach a useful multivariable definition for the derivative, it is useful to think geometrically.

In single-variable calculus, if $f'(x_o)$ exists, then the graph $y = f(x)$ must have a unique nonvertical tangent line at x_o. This means both that f is continuous at x_o and that $y = f(x)$ has no kinks or corners at x_o (see Figure 3.9). This attribute of the single-variable derivative is in fact much more helpful in defining the multivariable derivative. To make the presentation easier to visualize, suppose $n = 2$ and $m = 1$.

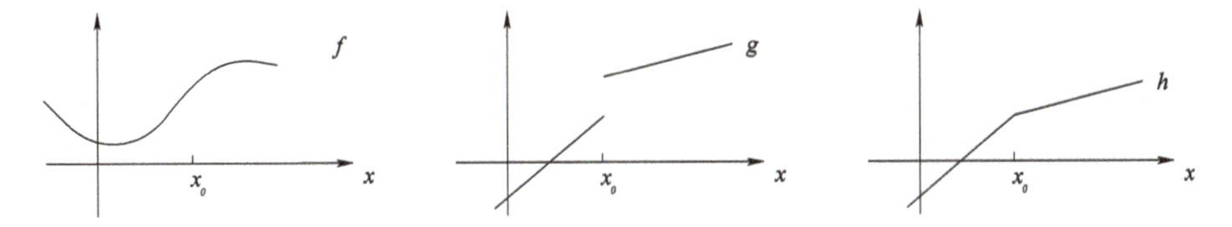

Figure 3.9: The graphs of three functions. Which functions are continuous at x_o? Which functions are differentiable at x_o?.

Definition. For some open domain $D \subset \mathbb{R}^2$, suppose that $f : D \to \mathbb{R}$. Then f is *differentiable* at $(x_o, y_o) \in D$ iff the graph of f (the surface $z = f(x,y)$ in $\mathbb{R}^3$) has a unique nonvertical tangent plane at $(x_o, y_o, f(x_o, y_o))$. If f is differentiable at every $(x,y) \in D$, then the surface $z = f(x,y)$ is *smooth* on D.

Remarks.

1. The above definition generalizes to other values of m and n, but it is then no longer possible to represent the tangent "plane" on a two-dimensional page.
2. This definition is intuitively appealing, but it suffers from one serious mathematical flaw: up to this point, we have not precisely defined what "a unique nonvertical tangent plane" is. Perhaps surprisingly, this definition is not too difficult to make, but does require some new concepts that we turn our attention to now.

3.3.1 Partial derivatives

Before we can meaningfully describe what it means to have a unique nonvertical tangent plane and the *entire* derivative, we must define and study an important *part* of the story: the partial derivative.

Definition. Suppose $\mathbb{R}^n$ is described by n variables, $x_1, x_2, \ldots, x_n$ (to begin with, these can be thought of as n rectangular coordinates, but in general, depending on n, they could be polar, cylindrical, spherical or any other well-defined coordinate system). As usual suppose $f : D \to \mathbb{R}$ where again $D \subset \mathbb{R}^n$ is an open domain for f in $\mathbb{R}^n$. The *partial derivative* of f with respect to x_i is

$$\frac{\partial f}{\partial x_i} := \lim_{\xi \to x_i} \frac{f(x_1, x_2, \ldots \xi, \ldots x_n) - f(x_1, x_2, \ldots x_i, \ldots x_n)}{\xi - x_i}$$

provided the above limit exists.

Remark. This definition explicitly contains the key idea behind the partial derivative: that all of the variables other than the one mentioned in the derivative notation are held constant. For those familiar with American football, this is similar to how the offensive must be set just before a play begins: all players except one must be stationary; one player only is allowed to be in motion. In the partial derivative, ξ is this moving player.

Notation. There may be no concept in mathematics with more notation than partial derivative. If $f : \mathbb{R}^2 \to \mathbb{R}$ is a real-valued function given by $z = f(x,y)$, then the following are all the equivalent:

$$\frac{\partial f}{\partial x} \equiv \frac{\partial z}{\partial x} \equiv f_x \equiv z_x \equiv \partial_x f \equiv \partial_x z$$

The fractional notation is an extension of the standard Leibniz derivative notation $\frac{df}{dx}$ and the subscript notation is an extension of the standard Newton prime (or dot) notation f'. The symbol "∂" is a Cyrillic italic "d" and is perhaps the only Cyrillic letter widely used in mathematics today.

Example 3.11. Suppose that $(x,y,z) \in D = \mathbb{R}^3$ and $f : D \to \mathbb{R}$ by $f(x,y,z) = 3x^2y^3 \sin z$. Then in this case:

$$\begin{aligned}
\frac{\partial f}{\partial x}(x,y,z) &= \lim_{\xi \to x} \frac{f(\xi,y,z) - f(x,y,z)}{\xi - x} \\
&= \lim_{\xi \to x} \frac{3\xi^2y^3 \sin z - 3x^2y^3 \sin z}{\xi - x} \\
&= 3y^3 \sin z \lim_{\xi \to x} \frac{\xi^2 - x^2}{\xi - x} \\
&= 3y^3 \sin z \lim_{\xi \to x} \frac{\cancel{(\xi - x)}(\xi + x)}{\cancel{\xi - x}} \\
&= 3y^3 \sin z(2x) = 6xy^3 \sin z
\end{aligned}$$

Notice that once the factors involving y and z are factored out, what remains is just a single-variable derivative—something that should be familiar from previous study. The two other partial derivatives, f_y and f_z can be computed in the same way, and one finds that they too are just what one gets by holding the other two variables constant and carrying out a standard single-variable differentiation with respect to the variable in question, either y or z. Indeed because of this, all of the standard rules of calculus (product rule, chain rule, etc.) apply in the usual way, and we can apply these rules to compute partial derivatives, rather than always computing the above limit. Still it should be kept in mind that in any situation where the standard rules do not apply (e. g., if one wants to compute a partial derivative of $f(x,y) = xy|xy|$ at $(x,y) = (0,0)$), one must fall back to computing this limit. But either by explicitly computing the limit, or by applying the standard rules, one can find that

$$\frac{\partial f}{\partial y}(x,y,z) = 9x^2y^2 \sin z$$

and that

$$\frac{\partial f}{\partial z}(x,y,z) = 3x^2y^3 \cos z\,.$$

3.3.2 Higher-order partial derivatives

As single-variable differentiation can be apply repeatedly to yield higher-order ordinary derivatives, so can partial derivatives . For example,

$$\frac{\partial^2 f}{\partial y^2} := \frac{\partial}{\partial y}\left(\frac{\partial f}{\partial y}\right)$$

again provided all of the limits implicit in this definition exist. In simple words, this is just the second partial of f with respect to y and it just means that f is differentiated with respect to y twice.

Example 3.12. Suppose that $f(t,x) = 2t\cot x$. Then $f_t(t,x) = 2\cot x$, and $f_{tt}(t,x) = 0$ for all (x,t), while $f_x(t,x) = -2t\csc^2 x$, and $f_{xx}(t,x) = 4t\csc^2 x\cot x$. One can also compute mixed partial derivative; in this case, one finds that

$$\frac{\partial^2 f}{\partial x \partial t} := \frac{\partial}{\partial x}\left(\frac{\partial f}{\partial t}\right) = -2\csc^2 x = \frac{\partial^2 f}{\partial t \partial x}.$$

Notation. There is a bit of an ambiguity here: for subscript notation, when the order matters, which order does one use? In this and most (but not all) other publications, the convention is

$$f_{xy} \equiv \frac{\partial}{\partial y}\left(\frac{\partial f}{\partial x}\right).$$

If it matters, the reader needs to be sure that a given publication follows this standard.

Example 3.12 leads to an obvious question: Are mixed partial derivatives always the same no matter which order the various derivatives are taken? It turns out that, perhaps surprisingly, the answer is no unless there is an important additional hypothesis.

Theorem 2. *Suppose that $f : D \subset \mathbb{R}^2 \to \mathbb{R}$ where D is an open domain for f. If $f \in C^2(D)$, meaning that f is twice continuously differentiable, so that it and all its partial derivatives up through second order (i. e., $f_x, f_y, f_{xx}, f_{xy}, f_{yx}$, and f_{yy}) exist and are continuous, then*

$$\frac{\partial^2 f}{\partial x \partial y}(x,y) = \frac{\partial^2 f}{\partial y \partial x}(x,y)$$

for all $(x,y) \in D$.

The proof of this theorem is not particularly difficult, but it is also not particularly instructive, so it is not included here; see Rudin [5, pp. 235–236].

3.3.3 Tangent planes and *unique* tangent planes

Suppose we are given a continuous surface $z = f(x,y)$ in $\mathbb{R}^3$. Can we find the equation for the tangent plane to this surface at some point $(x_o, y_o, z_o) = (x_o, y_o, f(x_o, y_o))$ on the surface? For example, can we find the equation for the tangent plane to the surface $z = 4 - x^2 - y^2$ at the point $(\sqrt{2}/2, \sqrt{2}/2, 3)$ as shown in Figure 3.10?

The most general linear equation in $\mathbb{R}^3$ (and thus the most general equation for a plane) is

$$Ax + By + Cz + D = 0.$$

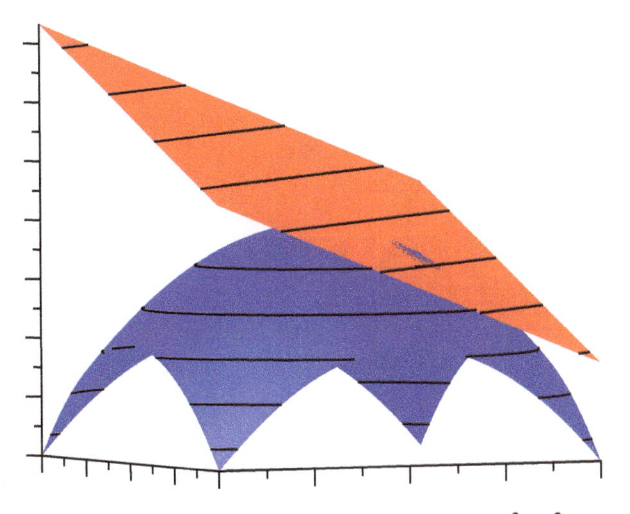

Figure 3.10: The tangent plane for the surface $z = 4 - x^2 - y^2$ at the point $(\sqrt{2}/2, \sqrt{2}/2, 3)$. What is the equation for this plane (see Section 3.2)?.

Any plane in $\mathbb{R}^3$ is defined by this equation for some choice of the constants A, B, C, and D. If $C = 0$, then the plane is vertical, that is, perpendicular to the x,y-plane, hence not the sort of plane we are looking for here. So assume that $C \neq 0$; one can then divide through by C and solve for z. The resulting equation has the form

$$z = d + ax + by$$

where $a := -A/C$, $b := -B/C$ and $d := -D/C$. The question now is how do we choose a, b, and d so that this becomes the equation of the tangent plane for the surface $z = f(x,y)$ at the point $(x_o, y_o, f(x_o, y_o))$.

First, let us consider what a must be. Consider the curve lying on the surface $z = f(x,y)$ passing through the point $(x_o, y_o, f(x_o, y_o))$ with y fixed at $y = y_o$ (*cf.* Figure 3.11). The equations for this curve is $z = f(x, y_o)$, $y = y_o$, and, of course, the slope of this curve is the derivative at (x_o, y_o): $f_x(x_o, y_o)$. So the line tangent to this curve (and, therefore, tangent to the surface) is the line passing through this point with slope $f_x(x_o, y_o)$ (again *cf.* Figure 3.11). Since a is the slope of the line $z = d + ax + by_o$ (recall that $y = y_o$ is here fixed),

$$a = f_x(x_o, y_o) \equiv \frac{\partial f}{\partial x}(x_o, y_o).$$

By similar reasoning, one finds that $b = f_y(x_o, y_o)$. To find d, notice that the tangent plane must also pass through the point $(x_o, y_o, f(x_o, y_o))$. Thus

$$z = f(x_o, y_o) = d + ax_o + by_o = d + \frac{\partial f}{\partial x}(x_o, y_o)x_o + \frac{\partial f}{\partial y}(x_o, y_o)y_o$$

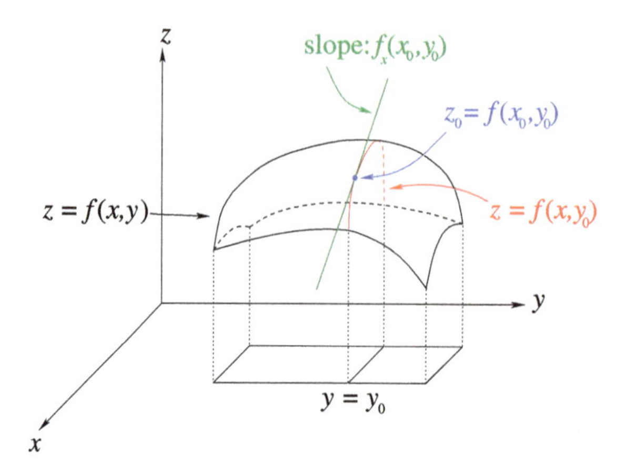

Figure 3.11: The partial derivative f_x at the point (x_o, y_o) is the slope of the tangent line with y_o fixed. The surface is $z = f(x,y)$. The curve $z = f(x, y_o)$ (in red) is the intersection of this surface and the plane $y = y_o$. The point of tangency (x_o, y_o, z_o) is in blue, while the tangent line is in green.

or

$$d = f(x_o, y_o) - \frac{\partial f}{\partial x}(x_o, y_o)x_o - \frac{\partial f}{\partial y}(x_o, y_o)y_o\,.$$

Using these values for a, b, and d, one finds the equation for the tangent plane:

$$\begin{aligned} z &= d + ax + by \\ &= \left[f(x_o, y_o) - \frac{\partial f}{\partial x}(x_o, y_o)x_o - \frac{\partial f}{\partial y}(x_o, y_o)y_o\right] + \left[\frac{\partial f}{\partial x}(x_o, y_o)\right]x + \left[\frac{\partial f}{\partial y}(x_o, y_o)\right]y \end{aligned}$$

or

$$\boxed{z = f(x_o, y_o) + \frac{\partial f}{\partial x}(x_o, y_o)\big(x - x_o\big) + \frac{\partial f}{\partial y}(x_o, y_o)\big(y - y_o\big)} \tag{3.1}$$

Example 3.13. Find the tangent plane to the surface $z = 4 - x^2 + y^3/3$ at the point $(1, 3, 12)$.

Answer. In this case, $f_x = -2x$ and $f_y = y^2$. So $f_x(1,3) = -2$ while $f_y(1,3) = 9$. Based on Equation 3.1, these imply that the tangent plane to $z = 4 - x^2 + y^3/3$ at $(1, 3, 12)$ is

$$\begin{aligned} z &= 12 - 2(x - 1) + 9(y - 3) \\ &= -2x + 9y - 13\,. \end{aligned}$$

The equation presented above (3.1) defines the unique nonvertical tangent plane for a surface $z = f(x,y)$, *provided* the surface actually has a tangent plane. That is, no other equation can give the tangent plane. But one might ask if there are any cases where all of the terms in this equation can be computed, but the surface does not actually have a tangent plane. Could a tangent plane not exist even though it would be unique if it did? The answer is "yes" and the following example presents such a case.

Example 3.14. Consider the surface that is given by

$$z = \begin{cases} \sin^2 6\theta & \pi/6 \le \theta \le \pi/3, \quad x > 0 \\ 0 & \text{elsewhere} \end{cases} .$$

This surface is shown graphically in Figure 3.12. Does this surface have a tangent plane at the origin $(0, 0, 0)$? The answer must be no, since the surface is not even continuous there. None the less, since $f(x, y) \equiv 0$ on both the x and y axes, each term in Equation (3.1) can be found, and indeed, $a = 0$, $b = 0$ and $c = 0$. So each of the partial derivatives can be found even though this surface has no tangent plane at the origin.

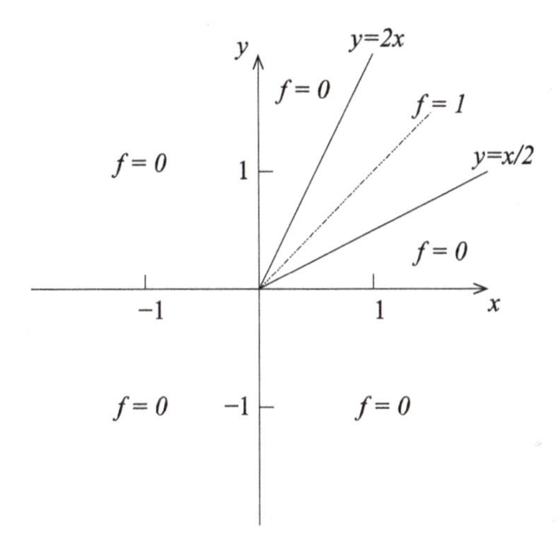

Figure 3.12: A surface with no tangent plane, but where the equation for the tangent plane can be computed. The surface is $z = f(x, y) = 0$ everywhere, *except* in the wedge between the two rays in the first quadrant. In this wedge, the surface varies smoothly, rising to a maximum value of $f = 1$ on the dashed line. What is the equation of the dashed line?

3.3.4 Existence of the tangent plane

The final example of the previous section gets at a sort of mathematical paradox: it is possible to compute the equation of a plane that is the only possible tangent plane, even though no tangent plane exists. In other words, uniqueness does not imply existence. How can one be certain that a tangent plane exists?

To answer this question, we must observe what works in the basic example above, but fails in the final example. The difference is that the slopes of tangent lines to curves in the first surface that pass near $(x_o, y_o, f(x_o, y_o))$ are all close to the slope of the tangent plane, while in the final example, the slopes of some curves passing near the origin are far from the zero slope of the plane $z = 0$. So in the first case, the tangent plane is a good approximation to the surface, both in the sense that points of the surface are close to the plane and in the sense that the slopes are close, while in the latter case, the plane is not at all a good approximation. The following definition makes this general idea mathematically exact.

Definition. For some open domain $D \subset \mathbb{R}^2$, suppose that $f : D \to \mathbb{R}$, and let $(x_o, y_o) \in D$. Then the surface $z = f(x,y)$ has *a unique nonvertical tangent plane* at $(x_o, y_o, f(x_o, y_o))$ provided that $f_x(x_o, y_o)$ and $f_y(x_o, y_o)$ both exist and that

$$\lim_{(x,y)\to(x_o,y_o)} \frac{f(x,y) - \left[f(x_o,y_o) + \frac{\partial f}{\partial x}(x_o,y_o)(x-x_o) + \frac{\partial f}{\partial y}(x_o,y_o)(y-y_o)\right]}{d((x,y),(x_o,y_o))} = 0$$

that is, the above limit exists and is zero.

Notice that quotient inside the limit is exactly the difference between the function and the only possible tangent plane, all divided by the distance between the base point (x_o, y_o) and the point (x,y) where the function and the plane are being evaluated. Hence for there to be a unique nonvertical tangent plane, the difference between the surface $z = f(x,y)$ and the potential tangent plane $z = f(x_o,y_o) + f_x(x_o,y_o)(x-x_o) + f_y(x_o,y_o)(y-y_o)$ must go to zero faster than the distance between (x,y) and (x_o,y_o) goes to zero.

Example 3.15. Show that the function defined by $f(x,y) = 4 - x^2 + y^3/3$ has a unique nonvertical tangent plane at $(1,3,12)$.

Answer. From Example 3.13 above, the only possible tangent plane is $z = -2x + 9y - 13$. So by the definition of unique nonvertical tangent plane:

$$\lim_{(x,y)\to(1,3)} \frac{4 - x^2 + y^3/3 - [-2x + 9y - 13]}{\sqrt{(x-1)^2 + (y-3)^2}} = 0$$

The expression inside the limit can be written in terms of shifted polar coordinates: define $\tilde{x} := x - 1$ and $\tilde{y} := y - 3$, and convert $(\tilde{x}, \tilde{y})$ to polar coordinates using $\tilde{x} = \tilde{r}\cos\tilde{\theta}$ and $\tilde{y} = \tilde{r}\sin\tilde{\theta}$. This shift is similar to the one in Example 3.10 above. Notice that this choice of coordinates is dictated by the distance expression in the denominator; this choice reduces the denominator to just $\tilde{r}$:

$$\begin{aligned}
\frac{4 - x^2 + y^3/3 - [-2x + 9y - 13]}{\sqrt{(x-1)^2 + (y-3)^2}} &= \frac{y^3 - 3x^2 - 27y + 6x + 51}{3\sqrt{(x-1)^2 + (y-3)^2}} \\
&= \frac{(\tilde{y}+3)^3 - 3(\tilde{x}+1)^2 - 27(\tilde{y}+3) + 6(\tilde{x}+1) + 51}{3\sqrt{\tilde{x}^2 + \tilde{y}^2}} \\
&= \frac{\tilde{y}^3 + 9\tilde{y}^2 - 3\tilde{x}^2}{3\sqrt{\tilde{x}^2 + \tilde{y}^2}} \\
&= \frac{\tilde{r}^3\sin^3\tilde{\theta} + 9\tilde{r}^2\sin^2\tilde{\theta} - 3\tilde{r}^2\cos^2\tilde{\theta}}{3\tilde{r}} \\
&= \frac{1}{3}\tilde{r}^2\sin^3\tilde{\theta} + 3\tilde{r}\sin^2\tilde{\theta} - \tilde{r}\cos^2\tilde{\theta}
\end{aligned}$$

Thus

$$\lim_{(x,y)\to(1,3)} \frac{4 - x^2 + y^3/3 - [-2x + 9y - 13]}{\sqrt{(x-1)^2 + (y-3)^2}} = \lim_{\tilde{r}\to 0} \left(\frac{1}{3}\tilde{r}^2 \sin^3\tilde{\theta} + 3\tilde{r}\sin^2\tilde{\theta} - \tilde{r}\cos^2\tilde{\theta}\right) = 0$$

since $\sin\theta$ and $\cos\theta$ are bounded between -1 and $+1$, and in this example, f has a unique nonvertical tangent plane at $(1, 3, 12)$.

3.3.5 Multivariable derivative

Everything discussed above deals with when a function has a tangent plane and when it is differentiable. But we have not yet stated what the derivative is in the multivariable setting—we need one more definition.

Definition. For some open domain $D \subset \mathbb{R}^n$, suppose that $\boldsymbol{f} : D \to \mathbb{R}^m$, and let $\boldsymbol{x}_o \in D$. If $\boldsymbol{f}$ is differentiable at $\boldsymbol{x}_o \in \mathbb{R}^n$, then the *derivative* of $\boldsymbol{f}$ at $\boldsymbol{x}_o$ is represented by the *Jacobian matrix* $\mathbf{D}\boldsymbol{f}(\boldsymbol{x}_o) \equiv J(\boldsymbol{f})(\boldsymbol{x}_o)$:

$$\mathbf{D}\boldsymbol{f}(\boldsymbol{x}_o) \equiv J(\boldsymbol{f})(\boldsymbol{x}_o) := \begin{bmatrix} \frac{\partial f_1}{\partial x_1}(\boldsymbol{x}_o) & \frac{\partial f_1}{\partial x_2}(\boldsymbol{x}_o) & \cdots & \frac{\partial f_1}{\partial x_n}(\boldsymbol{x}_o) \\ \frac{\partial f_2}{\partial x_1}(\boldsymbol{x}_o) & \frac{\partial f_2}{\partial x_2}(\boldsymbol{x}_o) & \cdots & \frac{\partial f_2}{\partial x_n}(\boldsymbol{x}_o) \\ \vdots & \vdots & & \vdots \\ \frac{\partial f_m}{\partial x_1}(\boldsymbol{x}_o) & \frac{\partial f_m}{\partial x_2}(\boldsymbol{x}_o) & \cdots & \frac{\partial f_m}{\partial x_n}(\boldsymbol{x}_o) \end{bmatrix}$$

The multivariable derivative has various meanings in various settings. Among the most important are the gradient of a multivariable function (when $f : \mathbb{R}^n \to \mathbb{R}$) and the derivative of a vector function (when $\boldsymbol{f} : \mathbb{R} \to \mathbb{R}^m$) that was discussed in the previous chapter. But in general for a vector field (when $\boldsymbol{f} : \mathbb{R}^n \to \mathbb{R}^m$) if f is differentiable, then the Jacobian matrix gives the linear approximation of $\boldsymbol{f}$ near $\boldsymbol{x}_o$:

$$\boldsymbol{f}(\boldsymbol{x}) \simeq \boldsymbol{f}(\boldsymbol{x}_o) + J(\boldsymbol{f})(\boldsymbol{x}_o) * (\boldsymbol{x} - \boldsymbol{x}_o)$$

where $*$ denotes matrix/vector multiplication and $\simeq$ should be read "approximately equal to" in a way that for the moment, we will not make exact. Notice that this multivariable linear approximation is reminiscent of the linear approximation of a single-variable function:

$$f(x) \simeq f(x_o) + f'(x_o)(x - x_o)$$

Example 3.16. Suppose that $\boldsymbol{f} : \mathbb{R}^3 \to \mathbb{R}^2$ by

$$\boldsymbol{f}(x, y, z) = \begin{bmatrix} f_1(x, y, z) \\ f_2(x, y, z) \end{bmatrix} = \begin{bmatrix} x^2 y \cos(z\pi) \\ 3xyz^2 \end{bmatrix}.$$

Please find the linear approximation of $\boldsymbol{f}$ near $(x_o, y_o, z_o) = (1, 3, -1)$, and describe what the approximation is.

Answer. Here,

$$f(x_o) = \begin{bmatrix} -3 \\ 9 \end{bmatrix}$$

and

$$J(f)(x_o) = \begin{bmatrix} 2xy\cos(z\pi) & x^2\cos(z\pi) & -\pi x^2 y\sin(z\pi) \\ 3yz^2 & 3xz^2 & 6xyz \end{bmatrix}\Bigg|_{(x_o,y_o,z_o)=(1,3,-1)}$$
$$= \begin{bmatrix} -6 & -1 & 0 \\ 9 & 3 & -18 \end{bmatrix}.$$

So the linear approximation in this case is

$$f(x) = f(x,y,z) \simeq \begin{bmatrix} -3 \\ 9 \end{bmatrix} + \begin{bmatrix} -6 & -1 & 0 \\ 9 & 3 & -18 \end{bmatrix} \begin{bmatrix} x-1 \\ y-3 \\ z+1 \end{bmatrix}$$

or just

$$f(x,y,z) \simeq \begin{bmatrix} -6x - y + 6 \\ 9x + 3y - 18z - 27 \end{bmatrix} \quad \text{near } (1,3,-1).$$

This linearization is two hyperplanes, one in each of the two dimensions of the range. These are hyperplanes rather than just planes in that each is a three-dimensional affine-linear subspace of four-dimensional space.

3.4 The chain rule in $\mathbb{R}^n$

At some point in single-variable calculus, students likely learn to remember the chain rule[4] as

$$\frac{dy}{dt} = \frac{dy}{dx}\frac{dx}{dt}$$

with the admonition not to think of this as cancellation of the dx "factors" on the right-hand side. Why cancellation is not the correct view becomes clear when one studies the multivariable chain rule (see Theorem 3 below).

4 Throughout this section, we assume that all the indicated derivatives exist. Also, the chain rule is likely the newest name given to a concept in calculus. Before about 1960, this rule was either unnamed or called the composition rule. After about 1960, it quickly became known as the chain rule in all calculus texts.

3.4.1 The basic chain rule

Suppose that $f, g, h : \mathbb{R}^2 \to \mathbb{R}$ are three differentiable functions with $w = f(x, y)$, $x = g(s, t)$ and $y = h(s, t)$. Then the chain rule implies that

$$\frac{\partial w}{\partial s} = \frac{\partial w}{\partial x}\frac{\partial x}{\partial s} + \frac{\partial w}{\partial y}\frac{\partial y}{\partial s}$$

and also

$$\frac{\partial w}{\partial t} = \frac{\partial w}{\partial x}\frac{\partial x}{\partial t} + \frac{\partial w}{\partial y}\frac{\partial y}{\partial t}.$$

The pattern for these chain rules is perhaps best remembered using the diagram in Figure 3.13. There, w is a function of x and y, which in turn are each functions of s and t, and derivatives of the top variable w with respect to either of the bottom variables s or t can be obtained by considering all of the paths through the middle variables x and y.

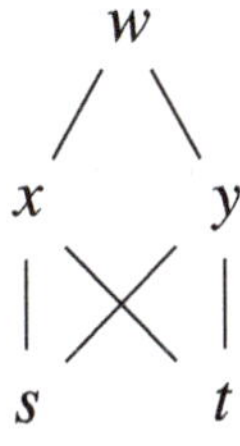

Figure 3.13: Dependency diagram for this first example. Notice that there are two paths from the top variable w to, for example, s at the bottom, one through x and the other through y. These correspond to the two terms in each of the chain rule expansions for $\partial w/\partial s$ and $\partial w/\partial t$.

The accuracy of these expressions is suggested by the following example.

Example 3.17. Suppose that $w = f(x, y) = e^x \cos y$, $x = g(s, t) = 2s + 5t$, and $y = h(s, t) = 3s - 7t$. Then substituting g and h into f, one finds that $w = e^{2s+5t}\cos(3s - 7t)$, and computing $\partial w/\partial s$ directly (recall that t is by definition a constant for this computation), one finds that

$$\frac{\partial w}{\partial s} = 2e^{2s+5t}\cos(3s - 7t) - 3e^{2s+5t}\sin(3s - 7t)$$

which of course is exactly the first expression above when all x and y are written in terms of s and t. Similarly,

$$\frac{\partial w}{\partial t} = 5e^{2s+5t}\cos(3s - 7t) + 7e^{2s+5t}\sin(3s - 7t)$$

which is the second expression.

Of course, an example does not prove that the expressions above are valid in general; a theorem is needed along with a clear proof.

Theorem 3. *For some open domain $U \subset \mathbb{R}^2$, suppose that $f : U \to \mathbb{R}$, and for some open domain $V \subset \mathbb{R}^2$, suppose that $\mathbf{g} : V \to \mathbb{R}^2$ with $\mathbf{g}(V) \subset U$. Suppose that $(s_o, t_o) \in V$, that $\mathbf{x}_o = \mathbf{g}(s_o, t_o) \in U$, and that $\mathbf{g}(s,t) = \langle g_1(s,t), g_2(s,t)\rangle$ when written componentwise. If f is differentiable at $\mathbf{x}_o$, and $\mathbf{g}$ is differentiable at (s_o, t_o), then $(f \circ \mathbf{g})(s,t) := f(g_1(s,t), g_2(s,t))$ is differentiable at (s_o, t_o), and if $w = (f \circ \mathbf{g})(s,t)$ with $x = g_1(s,t)$ and $y = g_2(s,t)$, then*

$$\frac{\partial w}{\partial s} = \frac{\partial w}{\partial x}\frac{\partial x}{\partial s} + \frac{\partial w}{\partial y}\frac{\partial y}{\partial s}$$

and

$$\frac{\partial w}{\partial t} = \frac{\partial w}{\partial x}\frac{\partial x}{\partial t} + \frac{\partial w}{\partial y}\frac{\partial y}{\partial t}$$

where all of the partial derivatives are evaluated at $\mathbf{x}_o$ and/or (s_o, t_o). This result can also be written as

$$\nabla(f \circ \mathbf{g})(s_o, t_o) = \nabla f(x_o, y_o) * J(\mathbf{g})(s_o, t_o)$$

where $J(\mathbf{g})(s_o, t_o)$ is the Jacobian matrix[5] *for the vector field $\mathbf{g}$ evaluated at (s_o, t_o), and $\nabla(f \circ \mathbf{g})(s_o, t_o) = \langle \frac{\partial w}{\partial s}, \frac{\partial w}{\partial t}\rangle$ and $\nabla f(x_o, y_o) = \langle \frac{\partial w}{\partial x}, \frac{\partial w}{\partial t}\rangle$ are* ***row vectors.***

The proof given here assumes that the partial derivatives are continuous; this assumption is not necessary for the theorem, but it does simplify the presentation and leads to a more-transparent proof using the mean-value theorem. For a proof not using this assumption, see Rudin [5, pp. 214–215].

Proof. First, consider $\partial w/\partial s$; from the definition of partial derivative, the variable t is held constant at t_o when this partial is computed. So using this definition, one finds that

$$\begin{aligned}
\frac{\partial w}{\partial s} &= \frac{\partial}{\partial s}[f(x_1(s,t), x_2(s,t))]\Big|_{(s_o,t_o)} \\
&= \lim_{s \to s_o} \frac{f(x_1(s,t_o), x_2(s,t_o)) - f(x_1(s_o,t_o), x_2(s_o,t_o))}{s - s_o} \\
&= \lim_{s \to s_o} \left[\frac{f(x_1(s,t_o), x_2(s,t_o)) - f(x_1(s_o,t_o), x_2(s,t_o))}{s - s_o}\right. \\
&\quad \left. + \frac{f(x_1(s_o,t_o), x_2(s,t_o)) - f(x_1(s_o,t_o), x_2(s_o,t_o))}{s - s_o}\right]
\end{aligned}$$

where $f(x_1(s_o, t_o), x_2(s, t_o))$ was subtracted and added in the numerator. Next, we apply the mean value theorem from single-variable calculus to both fractions in this sum. Recall that if F is a differentiable function of a single variable x, then $F(x) - F(x_o) =$

5 See Section 3.3.

$F'(c)(x - x_o)$ for some c between x and x_o. The same result works for multivariable functions, if all but one variable is fixed:

$$\frac{\partial w}{\partial s} = \lim_{s \to s_o}\left[\frac{\partial f}{\partial x_1}(c_1, x_2(s,t_o))\frac{x_1(s,t_o) - x_1(s_o,t_o)}{s - s_o} + \frac{\partial f}{\partial x_2}(x_1(s_o,t_o), c_2)\frac{x_2(s,t_o) - x_2(s_o,t_o)}{s - s_o}\right]$$

where for $k = 1, 2$, the value c_k is somewhere between $x_k(s_o,t_o)$ and $x_k(s,t_o)$. Finally, using the continuity of the two partial derivatives, one can take the limit through the sum:

$$\begin{aligned}\frac{\partial w}{\partial s} &= \frac{\partial f}{\partial x_1}(x_1(s_o,t_o), x_2(s_o,t_o)) \lim_{s \to s_o} \frac{x_1(s,t_o) - x_1(s_o,t_o)}{s - s_o} \\ &\quad + \frac{\partial f}{\partial x_2}(x_1(s_o,t_o), x_2(s_o,t_o)) \lim_{s \to s_o} \frac{x_2(s,t_o) - x_2(s_o,t_o)}{s - s_o} \\ &= \frac{\partial w}{\partial x}\frac{\partial x}{\partial s} + \frac{\partial w}{\partial y}\frac{\partial y}{\partial s}\end{aligned}$$

□

3.4.2 Several interesting extensions

The rest of this section expands on the basic multivariable chain rule, presenting by example several interesting extensions.

Example 3.18. Suppose $w = f(u,v)$, $u = g(x,y)$, $v = h(x,y,z)$, $x = \phi(r,s)$, $y = \xi(r,t)$ and $z = \zeta(s,t)$. If all of the functions are differentiable, use the chain rule to find $\frac{\partial w}{\partial r}$ and $\frac{\partial w}{\partial t}$.

Answer. The variable dependency diagram in Figure 3.14 may be helpful in determining this derivative. From this diagram, one can see that there are four paths from w to r and three from w to t, thus the chain rule expansion for $\frac{\partial w}{\partial r}$ has four terms, while the one for $\frac{\partial w}{\partial t}$ has three terms:

$$\frac{\partial w}{\partial r} = \frac{\partial w}{\partial u}\frac{\partial u}{\partial x}\frac{\partial x}{\partial r} + \frac{\partial w}{\partial u}\frac{\partial u}{\partial y}\frac{\partial y}{\partial r} + \frac{\partial w}{\partial v}\frac{\partial v}{\partial x}\frac{\partial x}{\partial r} + \frac{\partial w}{\partial v}\frac{\partial v}{\partial y}\frac{\partial y}{\partial r}$$

$$\frac{\partial w}{\partial t} = \frac{\partial w}{\partial u}\frac{\partial u}{\partial y}\frac{\partial y}{\partial t} + \frac{\partial w}{\partial v}\frac{\partial v}{\partial y}\frac{\partial y}{\partial t} + \frac{\partial w}{\partial v}\frac{\partial v}{\partial z}\frac{\partial z}{\partial t}$$

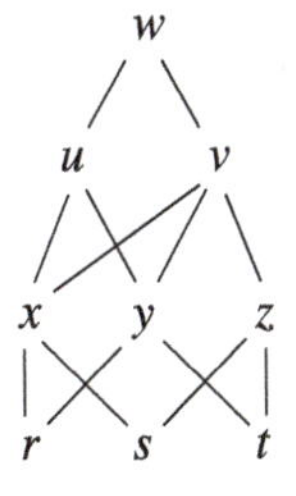

Figure 3.14: Dependency diagram for this second example. Why are there no lines between y and s, or between v and x?.

Example 3.19. In fluid dynamics and materials, one needs to distinguish between the partial derivative with respect to time and the total or material derivative with respect to time. Suppose that the velocity vector $\boldsymbol{v}$ for a small fluid or material element is a function of both the position of this element and time. If the motion is in three-dimensional space, then $\boldsymbol{v} = \boldsymbol{v}(t, x_1, x_2, x_3)$. But in addition, since the element is moving, its position is also a function of time. So there are functions ξ, η and ζ such that $x = \xi(t)$, $y = \eta(t)$ and $z = \zeta(t)$ meaning that $\boldsymbol{v} = \boldsymbol{v}(t, \xi(t), \eta(t), \zeta(t))$ Please find a chain rule expansion for the acceleration $\boldsymbol{a} := d\boldsymbol{v}/dt$, the total derivative of velocity with respect to time t.

Answer. As before in our discussion of vector functions, acceleration is defined to be the derivative of velocity with respect to all time dependencies, not just with respect to the first variable slot. The dependency diagram for velocity is given in Figure 3.15. Let $\boldsymbol{a} = \langle a_1, a_2, a_3 \rangle$ and $\boldsymbol{v} = \langle v_1, v_2, v_3 \rangle$. Noting that $dt/dt \equiv 1$, we can use the chain rule to compute each component of the acceleration:

$$\begin{aligned} a_i := \frac{dv_i}{dt} &= \frac{\partial v_i}{\partial t}\frac{dt}{dt} + \frac{\partial v_i}{\partial x_1}\frac{dx_1}{dt} + \frac{\partial v_i}{\partial x_2}\frac{dx_2}{dt} + \frac{\partial v_i}{\partial x_3}\frac{dx_3}{dt} \\ &= \frac{\partial v_i}{\partial t} + \left\langle \frac{\partial v_i}{\partial x_1}, \frac{\partial v_i}{\partial x_2}, \frac{\partial v_i}{\partial x_3} \right\rangle \cdot \left\langle \frac{dx_1}{dt}, \frac{dx_2}{dt}, \frac{dx_3}{dt} \right\rangle \\ &= \frac{\partial v_i}{\partial t} + \boldsymbol{v} \cdot \nabla v_i \end{aligned} \tag{3.2}$$

Here, $\langle dx_1/dt, dx_2/dt, dx_3/dt \rangle = \langle \dot{\xi}(t), \dot{\eta}(t), \dot{\zeta}(t) \rangle$ is the derivative of position with respect to time, thus it is the velocity vector $\boldsymbol{v}$, and ∇ is understood to be the vector of spacial derivatives. One often sees this component equation (3.2) written as a vector equation

$$\boldsymbol{a} = \frac{d\boldsymbol{v}}{dt} = \frac{\partial \boldsymbol{v}}{\partial t} + \boldsymbol{v} \cdot \nabla \boldsymbol{v}$$

but one must be careful to view the final term on the right as the dot product of $\boldsymbol{v}$ with ∇.

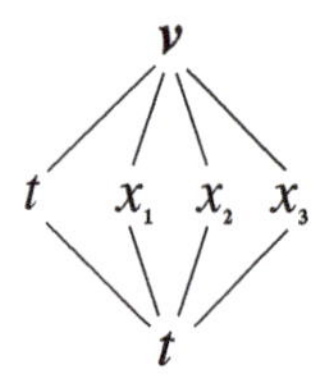

Figure 3.15: Dependency diagram for the velocity vector in fluid dynamics and materials.

3.4.3 Implicit partial differentiation

Suppose that $F : \mathbb{R}^n \to \mathbb{R}$ is a differentiable function, and consider the (hyper)surface in $\mathbb{R}^{n+1}$ defined by $F(x_1, x_2, \ldots, x_n) = 0$. How can one compute $\partial x_i/\partial x_j$ for $i \neq j$, and what does such a partial derivative represent? These questions will be answered here.

Theorem 4. *If $F : \mathbb{R}^n \to \mathbb{R}$ is a differentiable function and $F(x_1, x_2, \ldots, x_n) = 0$ is a (hyper)surface in $\mathbb{R}^n$. Then*

$$\frac{\partial x_i}{\partial x_j} = -\frac{F_{x_j}}{F_{x_i}}$$

provided that $F_{x_i} \neq 0$.

Remark. The presence of the negative sign makes clear that the idea of cancellation of "factors," even as a mnemonic, does not work for the multivariable chain rule.

Proof. Since $F(x_1, x_2, \ldots, x_n) = 0$ for all $(x_1, x_2, \ldots, x_n)$ on the hypersurface, both sides of this equation can be differentiated with respect to x_j; applying the chain rule yields

$$F_{x_1}\frac{\partial x_1}{\partial x_j} + F_{x_2}\frac{\partial x_2}{\partial x_j} + \cdots + F_{x_n}\frac{\partial x_n}{\partial x_j} = \frac{\partial}{\partial x_j}(0) = 0\,.$$

Now there are three values for the partial derivatives involving the variables: (1) $\partial x_k/\partial x_j = 0$ for $k \neq i, j$, because one holds x_k constant for $k \neq i, j$ when differentiating with respect to x_j; (2) $\partial x_j/\partial x_j = 1$; and (3) $\partial x_i/\partial x_j$ is the quantity to be found. These values imply that

$$F_{x_i}\frac{\partial x_i}{\partial x_j} + F_{x_j} = 0\,,$$

and if one solves for $\partial x_i/\partial x_j$, one finds the desired result. □

Example 3.20. Suppose a certain gas obeys the ideal gas law: $PV = nRT$ where P is pressure, V is volume, n is the number of gas molecules present, R is the universal gas constant, and T is temperature. How does volume change as pressure changes if all the other state variables are held constant?

Answer. Let $F(P, V, n, T) = PV - nRT = 0$ be our version of the ideal gas law. We wish to find $\partial V/\partial P$. Since $F_V = P$ and $F_P = V$, the desired partial derivative is $\partial V/\partial P = -V/P = -nRT/P^2$. Which of these two versions of the answer is preferred depends on which of the state variables one wishes to work with.

3.5 Directional derivatives

Previously, we have considered derivatives in certain directions, the coordinate directions. These are the partial derivatives. This section deals with derivatives in all directions—directional derivatives.

Definition. Suppose $f : D \to \mathbb{R}$ where as usual $D \subset \mathbb{R}^n$ is an open domain for f in $\mathbb{R}^n$. Suppose also that $\boldsymbol{u}$ is any unit vector in $\mathbb{R}^n$. The *directional derivative* of f in the direction of $\boldsymbol{u}$ at $\boldsymbol{x} \in D$ is

$$D_{\boldsymbol{u}} f(\boldsymbol{x}) := \lim_{h \to 0^+} \frac{f(\boldsymbol{x} + h\boldsymbol{u}) - f(\boldsymbol{x})}{h}$$

provided this limit exists.

Remark. The above definition for the directional derivative differs from that found in many texts in that it is one-sided. Many texts use $h \to 0$ rather than $h \to 0^+$ meaning that both the direction of $\boldsymbol{u}$ and the direction opposite $\boldsymbol{u}$ are both considered. The traditional definition has the advantage that it agrees with the definition of partial derivative when $\boldsymbol{u}$ lies in a coordinate direction. But the definition used here allows the directional derivative to exist in some meaningful cases where it would not if one was required to use a two-sided limit. Consider the following example.

Example 3.21. Compute the directional derivative for each direction for a certain cone opening downward with vertex at $(0, 0, 1)$:

$$z = f(x, y) = 1 - \sqrt{x^2 + y^2}$$

Answer. It is relatively easily to show that this surface has no unique tangent plane and is thus not differentiable at its vertex. But regarding the directional derivative, the limit to consider is

$$D_{\boldsymbol{u}} f(\mathbf{0}) := \lim_{h \to 0^+} \frac{f(\mathbf{0} + h\boldsymbol{u}) - f(\mathbf{0})}{h}$$

where $\boldsymbol{u} = \langle \cos\theta, \sin\theta \rangle$ is an easy way to represent the unit vector in any direction determined by $\theta \in [0, 2\pi)$. Hence

$$D_{\boldsymbol{u}} f(\mathbf{0}) := \lim_{h \to 0^+} \frac{-h\sqrt{\cos^2\theta + \sin^2\theta}}{h} = -1$$

independent of θ and the direction of $\boldsymbol{u}$. This is what one would expect standing on the pointed vertex of this cone: Every direction is the same and equally downhill.

One might think that having the same slope in each direction is in some sense the typical case—it is *not*. This is in fact the singular case where a generally smooth surface is not smooth at a specific point (in the previous example, this is the point of the cone), and thus it is not differentiable at a specific point. The generic case where a surface is smooth at and near a given point can be seen in the next example.

Example 3.22. For the paraboloid $z = f(x,y) = 4 - x^2 - y^2$ at the point $(\sqrt{2}/2, \sqrt{2}/2, 3)$, in which direction does the surface increase most rapidly? In which direction does it decrease most rapidly?

Answer. As in the previous example, the unit direction vector can be represented as $\boldsymbol{u} = \langle \cos\theta, \sin\theta \rangle$ for some $\theta \in [0, 2\pi)$. So the directional derivative is

$$\begin{aligned} D_{\boldsymbol{u}} f(\mathbf{0}) &= \lim_{h \to 0^+} \frac{f(\sqrt{2}/2 + h\cos\theta, \sqrt{2}/2 + h\sin\theta) - f(\sqrt{2}/2, \sqrt{2}/2)}{h} \\ &= \lim_{h \to 0^+} \frac{4 - (\sqrt{2}/2 + h\cos\theta)^2 - (\sqrt{2}/2 + h\sin\theta)^2 - 3}{h} \\ &= \lim_{h \to 0^+} \frac{-\sqrt{2}h\cos\theta - \sqrt{2}h\sin\theta - h^2}{h} \\ &= -\sqrt{2}(\cos\theta + \sin\theta) \end{aligned}$$

This time, the value of this limit definitely does depend on θ and, therefore, the direction. Noting the minus sign, one sees that the maximum value of the directional derivative occurs when $\theta = 5\pi/4$ and it minimum value occurs when $\theta = \pi/4$. All of this is exactly what one would expect on a paraboloid that opens downward: from this point (or any point) on the paraboloid, the direction of most rapid increase is the direction from the point toward the vertex, and the direction of most rapid decrease is opposite this direction.

The surface in the previous example along with its tangent plane at $(\sqrt{2}/2, \sqrt{2}/2, 3)$ is shown in Figure 3.10. Notice that on this tangent plane, the direction in the x, y-plane of most rapid increase on this tangent plane is opposite the direction in the x, y-plane of most rapid decrease on the tangent plane, and the plane is level in the perpendicular directions. This in fact is always true when a function f is differentiable, and there is a unique nonvertical tangent plane, as is stated in the next theorem.

Theorem 5. *Suppose that a function $f : D \subset \mathbb{R}^n \to \mathbb{R}$ is differentiable at some point $\boldsymbol{x} \in D$ for some open domain D. The directional derivative of f at $\boldsymbol{x}$ in the direction of a unit vector $\boldsymbol{u}$ is simply*

$$D_{\boldsymbol{u}} f(\boldsymbol{x}) = \nabla f(\boldsymbol{x}) \cdot \boldsymbol{u}$$

where $\nabla f := \langle f_{x_1}, f_{x_2}, \dots f_{x_n} \rangle$ is the gradient*: the vector of first partial derivatives (the nth component is the nth first partial derivative of f).*

Proof. Since the function f is differentiable at $\boldsymbol{x}$, the surface (or hypersurface) $z = f(x_1, x_2, \dots, x_n)$ has a unique, nonvertical tangent plane (or hyperplane), and thus

$$\lim_{\xi \to x} \frac{f(\boldsymbol{\xi}) - \left[f(\boldsymbol{x}) + \frac{\partial f}{\partial x_1}(\boldsymbol{x})(\xi_1 - x_1) + \frac{\partial f}{\partial x_2}(\boldsymbol{x})(\xi_2 - x_2) + \cdots + \frac{\partial f}{\partial x_n}(\boldsymbol{x})(\xi_n - x_n)\right]}{d(\boldsymbol{\xi}, \boldsymbol{x})} = 0$$

or when the terms are rearranged:

$$\lim_{\xi \to x} \frac{f(\boldsymbol{\xi}) - f(\boldsymbol{x})}{d(\boldsymbol{\xi}, \boldsymbol{x})} = \lim_{\xi \to x} \frac{\frac{\partial f}{\partial x_1}(\boldsymbol{x})(\xi_1 - x_1) + \frac{\partial f}{\partial x_2}(\boldsymbol{x})(\xi_2 - x_2) + \cdots + \frac{\partial f}{\partial x_n}(\boldsymbol{x})(\xi_n - x_n)}{d(\boldsymbol{\xi}, \boldsymbol{x})}$$

Then the directional derivative by definition is

$$D_{\boldsymbol{u}} f(\boldsymbol{x}) := \lim_{h \to 0^+} \frac{f(\boldsymbol{x} + h\boldsymbol{u}) - f(\boldsymbol{x})}{h}$$

which means that $\boldsymbol{\xi} = \boldsymbol{x} + h\boldsymbol{u}$ in the differentiability definition above. Thus $d(\boldsymbol{\xi}, \boldsymbol{x}) = h$ because $\boldsymbol{u}$ is a unit vector, and hence

$$\begin{aligned} D_{\boldsymbol{u}} f(\boldsymbol{x}) &= \lim_{h \to 0^+} \frac{f(\boldsymbol{\xi}) - f(\boldsymbol{x})}{d(\boldsymbol{\xi}, \boldsymbol{x})} \\ &= \lim_{h \to 0^+} \frac{\frac{\partial f}{\partial x_1}(\boldsymbol{x})(\xi_1 - x_1) + \frac{\partial f}{\partial x_2}(\boldsymbol{x})(\xi_2 - x_2) + \cdots + \frac{\partial f}{\partial x_n}(\boldsymbol{x})(\xi_n - x_n)}{d(\boldsymbol{\xi}, \boldsymbol{x})} \\ &= \lim_{h \to 0^+} \frac{\nabla f(\boldsymbol{x}) \cdot \langle \xi_1 - x_1, \xi_2 - x_2, \ldots \xi_n - x_n \rangle}{h} \\ &= \lim_{h \to 0^+} \frac{\nabla f(\boldsymbol{x}) \cdot (\not{h}\boldsymbol{u})}{\not{h}} \\ &= \nabla f(\boldsymbol{x}) \cdot \boldsymbol{u} \end{aligned}$$

□

Theorem 5 has an immediate important implication whose proof is a direct application of this theorem.

Corollary 2. *Suppose that a function $f : D \subset \mathbb{R}^n \to \mathbb{R}$ is differentiable at some point $\boldsymbol{x} \in D$ for some open domain D. Then the gradient $\nabla f(\boldsymbol{x})$ determines the direction of most rapid increase and decrease for the function f and the surface $z = f(x_1, x_2, \ldots, x_n)$ at $\boldsymbol{x} \in D$: Provided that $|\nabla f(\boldsymbol{x})| \neq 0$, then*

- $\boldsymbol{u} = \frac{\nabla f(\boldsymbol{x})}{|\nabla f(\boldsymbol{x})|}$ *is the unit vector pointing in the direction of most rapid increase for f, and the rate of increase in this direction is $|\nabla f(\boldsymbol{x})|$.*
- $\boldsymbol{u} = -\frac{\nabla f(\boldsymbol{x})}{|\nabla f(\boldsymbol{x})|}$ *is the unit vector pointing in the direction of most rapid decrease for f, and the rate of decrease in this direction is $-|\nabla f(\boldsymbol{x})|$.*
- *Any $\boldsymbol{u}$ with $\boldsymbol{u} \perp \nabla f(\boldsymbol{x})$ points in a direction where the directional derivative is zero, and hence f is not changing.*

Remark. Notice that $\nabla f(\boldsymbol{x})$ is a vector in $\mathbb{R}^n$ while $z = f(x_1, x_2, \ldots, x_n)$ is a surface in $\mathbb{R}^{n+1}$. So, for example, if $n = 2$ and $z = f(x, y)$ is thought of as a hillside with someone standing at $(x, y, f(x, y))$, then the gradient $\nabla f(x, y)$ points horizontally as a compass does. But unlike a standard compass that points north, this gradient vector points in the horizontal direction corresponding to walking uphill most steeply.

Example 3.23. For the surface $z = f(x,y) = \sin(x)\cos(y)$ at the point $(\pi/4, \pi/4, 1/2)$, in which direction does the surface decrease most rapidly? In which direction is the surface level?

Answer. According to Corollary 2, the direction of steepest decrease is that opposite the gradient:

$$-\nabla f(\pi/4,\pi/4) = -\langle \cos(\pi/4)\cos(\pi/4), -\sin(\pi/4)\sin(\pi/4)\rangle = \langle -1/2, 1/2\rangle = \frac{1}{2}\langle -1, 1\rangle$$

or simply the direction of $\langle -1, 1\rangle$. The level directions are those perpendicular to $\langle -1, 1\rangle$, so either $\langle 1, 1\rangle$ or $\langle -1, -1\rangle$.

Exercises 3

3.1. Consider two points in $\mathbb{R}^5$: $\mathbf{x} = (3, 0, -2, 1, \pi)$ and $\mathbf{y} = (7, -1, 2, 1, -\pi)$.

(a) What is the distance between $\mathbf{x} = (3, 0, -2, 1, \pi)$ and $\mathbf{y} = (7, -1, 2, 1, -\pi)$?

Answer. $\sqrt{33 + 4\pi^2}$

(b) Which vector $\mathbf{v}$ has its tail at $\mathbf{x}$ and its head at $\mathbf{y}$, and what is its length?

Answer. $\langle 4, -1, 4, 0, -2\pi\rangle$, $\sqrt{33 + 4\pi^2}$

3.2. Please find the equation satisfied by all points equidistant from $(3, -1, 1)$ and $(1, 2, 5)$. **Hint:** Equate the distance squared of any point (x, y, z) satisfying the equation and each of the two given points.

3.3. Please describe the values of α and β for which the distance between the points $\mathbf{x} = (\alpha, 1, \alpha)$ and $\mathbf{y} = (\beta, \beta, 1)$ equals 1?

Answer. The circle which is the intersection of the sphere $(\alpha - 1)^2 + (\beta - 1)^2 + \gamma^2 = 1$ and the plane $\gamma = \alpha - \beta$ in α, β, γ-space.

3.4. Calculate the path-dependent limit for the railroad underpass surface (Example 3.3) when $y = -2x$ and $x < 0$; show that for this path, the value of the path-dependent limit is $-2/5$.

3.5. For each of the following, please compute the limit, or explain why it does not exist. In some cases, considering separate paths will be helpful. In other cases, polar coordinates, factoring/cancellation or a substitution will be helpful. Also computer plotting software may be helpful:

(a) $\lim_{(x,y)\to(4,3)} x^2 + 2xy$

(b) $\lim_{\substack{(x,y)\to(0,0)\\ y\neq\pm x}} \frac{x-y}{x^2-y^2}$

(c) $\lim_{(x,y)\to(0,0)} e^{x+y}$

(d) $\lim_{\substack{(x,y)\to(1,1)\\ y\neq x}} \frac{x^6-y^6}{x^2-y^2}$

(e) $\lim_{(x,y)\to(0,0)} \frac{x^2-y^2}{x^2+y^2}$

(f) $\lim_{(x,y)\to(0,0)} \frac{x+y}{\sqrt{x^2+y^2}}$

(g) $\lim_{\substack{(x,y)\to(1,3)\\ y\neq 3x}} \frac{27x^3-y^3}{9x^2-y^2}$

(h) $\lim_{(x,y)\to(0,0)} \frac{\sin 4(x^2+y^2)}{x^2+y^2}$

3.6. The expressions below are the same as those in the limits in the previous exercise. In each case, is the expression continuous at (x_o, y_o), or is it possible to extend the definition of the expression so that it is continuous at (x_o, y_o)? If such an extension is possible, what value should the expression be defined as to make it continuous at (x_o, y_o).

(a) $x^2 + 2xy$ at $(x_o, y_o) = (4, 3)$

(b) $\frac{x-y}{x^2-y^2}$ at $(x_o, y_o) = (0, 0)$

(c) e^{x+y} at $(x_o, y_o) = (0, 0)$

(d) $\frac{x^6-y^6}{x^2-y^2}$ at $(x_o, y_o) = (1, 1)$

(e) $\frac{x^2-y^2}{x^2+y^2}$ at $(x_o, y_o) = (0, 0)$

(f) $\frac{x+y}{\sqrt{x^2+y^2}}$ at $(x_o, y_o) = (0, 0)$

(g) $\frac{27x^3-y^3}{9x^2-y^2}$ at $(x_o, y_o) = (1, 3)$

(h) $\frac{\sin 4(x^2+y^2)}{x^2+y^2}$ at $(x_o, y_o) = (0, 0)$

3.7. Please prove the remaining portions of Proposition 4. For the limit of $\boldsymbol{f} - \boldsymbol{g}$, follow the proof give in the text for the limit of $\boldsymbol{f} + \boldsymbol{g}$. For the limit of the scalar product, let $\alpha_L := \lim_{\boldsymbol{x}\to\boldsymbol{x}_o} \alpha(\boldsymbol{x}), \boldsymbol{f}_L := \lim_{\boldsymbol{x}\to\boldsymbol{x}_o} \boldsymbol{f}(\boldsymbol{x})$, and notice that

$$\begin{aligned}|\alpha(\boldsymbol{x})\boldsymbol{f}(\boldsymbol{x}) - \alpha_L\boldsymbol{f}_L| &= |\alpha(\boldsymbol{x})\boldsymbol{f}(\boldsymbol{x}) - \alpha_L\boldsymbol{f}(\boldsymbol{x}) + \alpha_L\boldsymbol{f}(\boldsymbol{x}) - \alpha_L\boldsymbol{f}_L| \\ &\le |\alpha(\boldsymbol{x}) - \alpha_L||\boldsymbol{f}(\boldsymbol{x})| + |\alpha_L||\boldsymbol{f}(\boldsymbol{x}) - \boldsymbol{f}_L|\,.\end{aligned}$$

Prove the result for the dot product with $m = 2$ in a similar manner, but skip the proof for the cross product (it is just too messy!).

3.8. Please extend the function $f(x,y) = (e^{xy^2} - 1)/x$ continuously to the line $x = 0$.

3.9. Suppose functions $f : D \subset \mathbb{R}^2 \to \mathbb{R}$ are given by each of the expressions listed below, with D being all of $\mathbb{R}^2$ where the denominator is not zero. Following Example 3.10 above, decide whether or not each function can it be extended to be continuous for all of $\mathbb{R}^2$?

(a) $f(x,y) = \frac{x^2y^2+6x^2y-4xy^2-24xy+14x^2+9y^2-56x+54y+101}{x^2+y^2-4x+6y+13}$

(b) $f(x,y) = \frac{9x^2-16y^2}{x^2+y^2-25}$

Answer. (b) This rational function is continuous except on the circle where the denominator is zero. There are four points where both the numerator and denominator are zero, but the limits approaching these points all depend on the direction of approach and thus do not exist.

3.10. Characterize the discontinuity of

$$f(x,y) = \frac{1}{y-x}$$

where it is not continuous (and perhaps not defined) on the x,y-plane. Is the discontinuity removable, a jump, a pole, or some combination of these?

3.11. For each of the following expressions, please find the first and second partial derivatives:

$$\frac{\partial}{\partial x}, \quad \frac{\partial}{\partial y}, \quad \frac{\partial^2}{\partial x^2}, \quad \frac{\partial^2}{\partial x \partial y} \quad \text{and} \quad \frac{\partial^2}{\partial y^2}$$

(a) $x^2 + 2xy$

(b) $x^2 y \sin(xy^2)$

(c) $xye^{x^2+y^2}$

(d) $\frac{x+y}{\sqrt{x^2+y^2}}$

3.12. For most functions in calculus, mixed partials are equal regardless of the order in which the derivatives are computed. This is a famous example where the mixed partials are *not* equal. Consider $f : \mathbb{R}^2 \to \mathbb{R}$ given by

$$f(x,y) = \begin{cases} \frac{xy(x^2-y^2)}{x^2+y^2} & (x,y) \neq (0,0) \\ 0 & (x,y) = (0,0) \end{cases}$$

First compute $f_x(x,y)$ and $f_y(x,y)$ using the standard rules for computing derivatives (the quotient rule, the chain rule, etc.). Then show that $f_x(0,0) = 0$ and $f_y(0,0) = 0$ using the definition of the derivative for these partial derivatives (the standard rules do not apply at $(0,0)$), and finally show that $f_{xy}(0,0) = -1$ while $f_{yx}(0,0) = 1$ again by applying the definition of the derivative. Finally show that f_{xy} is not continuous at $(0,0)$.

3.13. Please find an equation for the tangent plane to the surface $z = x \sin y$ at $(2, \pi/6, 1)$.

3.14. For $\boldsymbol{f}(x,y,z) = \langle xy^2,\ 2xy^2 \sin z \rangle$,
(a) Please find Jacobian matrix $J\boldsymbol{f}(x,y,z)$.
(b) Evaluate $J\boldsymbol{f}(-1,1,\pi/3)$.
(c) Compute the linear approximation of $\boldsymbol{f}$ near $(-1,1,\pi/3)$.

3.15. Please explain why the coefficient of y in the equation of the tangent plane to the surface $z = f(x,y)$ at the point $(x_o, y_o, f(x_o, y_o))$ must be $f_y(x_o, y_o)$. Draw a diagram similar to that in Figure 3.11 but now illustrating $f_y(x_o, y_o)$ as the slope of the tangent curve $z = f(x_o, y)$.

3.16. Show that the surface $z = f(x,y) = xy^2$ has a unique nonvertical tangent plane at $(1,1,1)$ using the definition. **Hint:** Follow Example 3.15.

3.17. If $F : D \subset \mathbb{R}^3 \to \mathbb{R}$, what limit must be zero if the hypersurface $w = F(x,y,z)$ is to have a unique nonvertical tangent hyperplane at (x_o, y_o, z_o)? **Hint:** Consider the definition of unique nonvertical tangent plane.

3.18. Suppose the definition of the directional derivative is changed from a one-sided limit to a two-sided limit:

$$\tilde{D}_{\boldsymbol{u}} f(\boldsymbol{x}) := \lim_{h \to 0} \frac{f(\boldsymbol{x} + h\boldsymbol{u}) - f(\boldsymbol{x})}{h}$$

provided this limit exists. Show that $\tilde{D}_{\boldsymbol{u}}f(\mathbf{0})$ does not exist for any direction for the cone $z = f(x,y) = 1 - \sqrt{x^2 + y^2}$. (This is the cone discussed in the first example in the directional derivative section.)

3.19. Suppose $w = f(u,v) = u + v^2$, $u = g(x,y,z) = x^2 + 2y + 4z$, $v = h(x,z) = xz$, $x = \phi(r,s) = 2r + 3s$, $y = \xi(r,t) = r\cos t$, and $z = \zeta(s,t) = 4s + 3t$. Please find $\partial w/\partial x$ and $\partial w/\partial t$. **Hint:** A dependency diagram will help.

3.20. Following the proof of Theorem 3 for $\partial w/\partial s$, please obtain a proof for the chain rule expansion for $\partial w/\partial t$.

3.21. Suppose $f(x,y,z) = e^y(x + \sin(z))$.
(a) Please compute $\frac{\partial f}{\partial x}$ and $\frac{\partial f}{\partial z}$.
(b) Please compute $\frac{df}{dx}$ and $\frac{df}{dz}$.
(c) Please compute $\frac{df}{dx}$ for $f(x,x^2,x^3)$, both directly using substitution and by using the chain rule.

Answer. (b) $\frac{df}{dx} = \frac{\partial f}{\partial x} + \frac{\partial f}{\partial y}\frac{\partial y}{\partial x} + \frac{\partial f}{\partial z}\frac{\partial z}{\partial x} = e^y + e^y(x + \sin(z))\frac{\partial y}{\partial x} + e^y\cos(z)\frac{\partial z}{\partial x}$

3.22. For $x^2(1 + \sqrt{y + z^2})e^{xy^2z^2} = 1$, please compute $\partial z/\partial x$ implicitly.

3.23. An alternative to the ideal gas law is the (more elaborate) van der Waals gas law: $(P + n^2a/V^2)(V - nb) = nRT$ where a and b are constants that depend on the specific gas. Please find $\partial V/\partial P$ relative to the van der Waals gas law.

3.24. One of the interesting results in thermodynamics is that

$$\frac{\partial V}{\partial T}\frac{\partial P}{\partial V}\frac{\partial T}{\partial P} = -1$$

regardless of which gas law $F(P,V,n,T) = 0$ is used. Please prove this using implicit differentiation.

3.25. Prove that the basic multivariable chain rule implies the expansion in Example 3.18.

3.26. For the surface $z = f(x,y) = x^2+y^2$, please find the directional derivative $D_{\boldsymbol{u}}f(1,2)$ in the direction given by the vector $\langle 2,1\rangle$.

Answer. $D_{\boldsymbol{u}}f(1,2) = 8\sqrt{5}/5$

3.27. For the surface $z = f(x,y) = \sin(x)\cos(y)$ at the point $(\pi/4, \pi/4, 1/2)$, what is the maximum rate of increase in f, that is, what is the maximum value of the directional derivative? In which direction does this increase occur?

3.28. For the surface $z = f(x,y) = xy^2e^{xy}$ in which directions from the origin is the directional derivative equal to half of its maximum value?

4 Implications of multivariable derivatives

Several important implications of multivariable derivatives are covered in this section. Before discussing these, however, we introduce one more basic topic: *level* curves and surfaces.

4.1 Level curves, level surfaces

There are many ways that various curves (or surfaces) can relate to each other. One of the most important is that of being level curves (or level surfaces).

Definition. Given a function $f : D \to \mathbb{R}$ for some domain $D \subset \mathbb{R}^2$, the *level curves* or *contours* for f are the curves of the form $f(x,y) = c$ for some constant c in the range of f.

This definition may seem complicated, but a simple example should clarify the concept.

Example 4.1. Suppose that $f(x,y) = x^2 + y^2$. The level curves for this function f are circles of the form $x^2 + y^2 = c$ where in this case $\sqrt{c}$ is the radius of the circle provided of course that $c > 0$. The one special case is $c = 0$ which is a single point, the origin. It can be thought of as a degenerate circle in this case. This family of level surfaces is shown in Figure 4.1.

The concept of level curves can be extended to three-dimensional space in the obvious way.

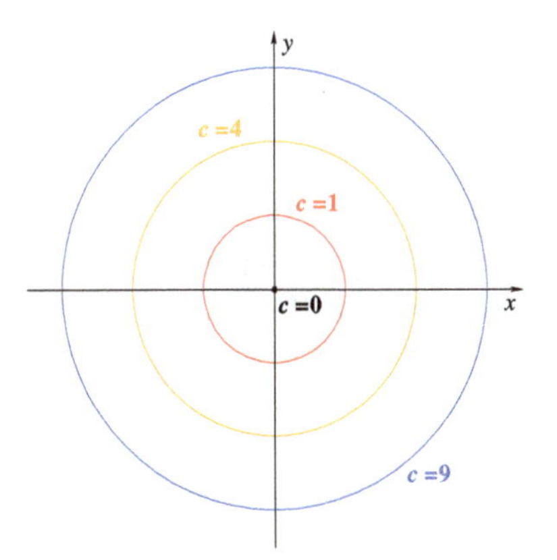

Figure 4.1: Level curves for $f(x,y) = x^2 + y^2$. Here, level curves are circles; the constant values shown are $c_0 = 0$, $c_1 = 1$, $c_2 = 4$, and $c_3 = 9$.

https://doi.org/10.1515/9783110660609-004

Definition. Given a function $F : D \to \mathbb{R}$ for some domain $D \subset \mathbb{R}^3$, the *level surfaces* for F are the surfaces of the form $F(x,y,z) = c$ for some constant c in the range of F.

In principle, one can extend this concept to even higher dimensions.

Example 4.2. Consider the function F defined as

$$F(x,y,z) = \frac{x^2}{\alpha^2} + \frac{y^2}{\beta^2} + \frac{z^2}{\gamma^2}$$

where $\alpha, \beta, \gamma > 0$ are constants. Notice that $F(x,y,z) = c^2$ is a family of ellipsoids with principal axes $c\alpha$, $c\beta$, and $c\gamma$.

Example 4.3. Consider the function G defined as $G(x_1, x_2, x_3, x_4) = x_4 - x_3^2 - x_2^2 - x_1^2$. Then $G(x_1, x_2, x_3, x_4) = c$ is a family spherical-hyperparaboloids opening upward in x_4. For each fixed value of $x_4 > c$, each three-dimensional surface in (x_1, x_2, x_3) is a sphere with radius $R = \sqrt{x_4 - c}$.

4.2 The gradient ∇F for the surface $F(x,y,z) = 0$

Previously, we discussed the role and significance of the gradient ∇f for the surface $z = f(x,y)$, that is, when one can explicitly solve for z in terms of the other two variables x and y. Now consider the case when one cannot (or perhaps has not) solved for any of the three variables x, y and z. So here $F : \mathbb{R}^3 \to \mathbb{R}$, and we consider the level surface $F(x,y,z) = 0$.[1] The big question to consider now is "Where is the gradient ∇F relative to the surface $F(x,y,z) = 0$?" This is both a very interesting result in its own right, and a very good review of many of the key concepts that have been covered so far.

Theorem 6. *Suppose that $F : \mathbb{R}^3 \to \mathbb{R}$ is differentiable. Let (x_o, y_o, z_o) be any point on the surface $F(x,y,z) = 0$. Then the gradient vector $\nabla F(x_o, y_o, z_o)$ is perpendicular (normal) to this surface at (x_o, y_o, z_o).*

This result is depicted in Figure 4.2.

Proof. Suppose that $\gamma = (x_1, x_2, x_3)$ is a parameterization of a smooth curve that lies on the surface and passes through the point (x_o, y_o, z_o) when $t = t_o$. So $F(x_1(t), x_2(t), x_3(t)) = 0 \; \forall t$, and $\gamma(t_o) = (x_o, y_o, z_o)$. Define the vector function $\boldsymbol{x}(t) := \langle x_1(t), x_2(t), x_3(t)\rangle$. Since F is differentiable and $F(x_1(t), x_2(t), x_3(t)) = 0$ for all t (the curve is on the surface), it makes sense to differentiate both sides of this identity with respect to t:

$$\frac{d}{dt} F(x_1(t), x_2(t), x_3(t)) = \frac{d}{dt}\big(0\big) = 0\,.$$

1 In fact, $F(x,y,z) = c$ works equally well for any constant c in the range of F.

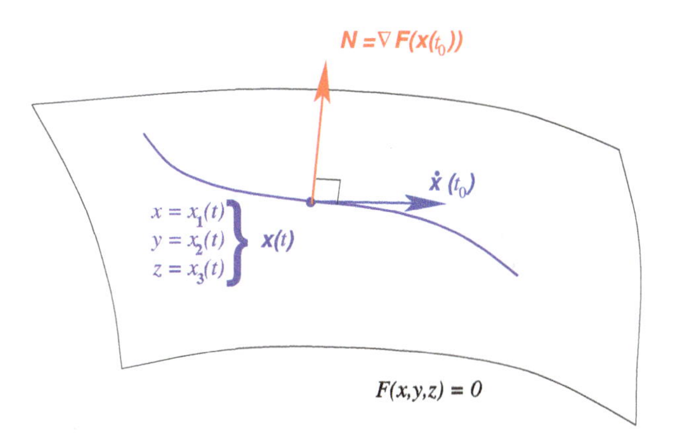

Figure 4.2: The gradient $\nabla F(x_o, y_o, z_o)$ is perpendicular to any curve lying in the surface, and thus perpendicular to the surface itself.

Applying the chain rule to the lefthand side, one finds that

$$\begin{aligned}\frac{d}{dt}F(x_1(t), x_2(t), x_3(t)) &= F_x(x_1(t), x_2(t), x_3(t))\dot{x}_1(t) + F_y(x_1(t), x_2(t), x_3(t))\dot{x}_2(t)\\ &\quad + F_z(x_1(t), x_2(t), x_3(t))\dot{x}_3(t)\\ &= \nabla F(x_1(t), x_2(t), x_3(t)) \cdot \dot{\boldsymbol{x}}(t)\,.\end{aligned}$$

Combining these two expressions for $\frac{d}{dt}F(x_1(t), x_2(t), x_3(t))$, one finds that at $t = t_o$

$$\nabla F(x_o, y_o, z_o) \cdot \dot{\boldsymbol{x}}(t_o) = 0\,.$$

Now recall what the dot product of two vectors being zero implies: they must be perpendicular. And since $\dot{\boldsymbol{x}}(t_o)$ is tangent to the curve defined by $\boldsymbol{x}(t)$, the gradient $\nabla F(x_o, y_o, z_o)$ must also be perpendicular to the curve. Now since this calculation applies to any curve lying in the surface, $\nabla F(x_o, y_o, z_o)$ must be perpendicular (or normal) to the surface. □

Example 4.4. For $F(x, y, z) = 2x^3 + 3xy^2 - xyz + 5yz^2 - 9$, please find the equation of the plane tangent to the surface $F(x, y, z) = 0$ at $(1, 1, 1)$.

Answer. First, it is important to confirm that $F(1, 1, 1) = 0$, but indeed this is the case. Applying Theorem 6, one finds that $\boldsymbol{N} = \nabla F(1, 1, 1) = \langle 8, 10, 9\rangle$. So the equation of the tangent plane is just

$$8(x - 1) + 10(y - 1) + 9(z - 1) = 0 \quad \Leftrightarrow \quad 8x + 10y + 9z = 27$$

Remark. The previous theorem tells us that ∇F is perpendicular to the surface $F(x, y, z) = 0$. The same sort of result is true in other dimensions: in $\mathbb{R}^2$, if $f(x, y) = 0$ is a curve, then ∇f is perpendicular to this curve (see Exercise 4.7); in $\mathbb{R}^n$, if $F(x_1, x_2, \ldots, x_n) = 0$ is a hypersurface, then ∇F is perpendicular to this hypersurface.

4.3 Maximums and minimums for continuous functions on closed and bounded domains

Toward the middle of the 19th century, almost any mathematician would have believed that a continuous function on a closed and bounded set would achieve its absolute minimum and maximum values, but none of them would have known exactly how to prove it. The key concept which finally allowed a full, rigorous proof was compactness. The details of why compactness is important and a proof of the result is beyond the scope of this text (for details see, e. g., Rudin [5] or Rosenlicht [4]), but the following theorem is the foundation of what is presented below.

Theorem 7. *Suppose that a function $F : D \subset \mathbb{R}^n \to \mathbb{R}$ is continuous on some closed and bounded subset $Q \subset D$. Then F achieves both its maximum and minimum values on Q, specifically, $\exists\, \boldsymbol{x}_{\mathrm{M}} \in Q$ and $\boldsymbol{x}_{\mathrm{m}} \in Q$ such that $\forall \boldsymbol{x} \in Q$, $F(\boldsymbol{x}_{\mathrm{m}}) \leq F(\boldsymbol{x}) \leq F(\boldsymbol{x}_{\mathrm{M}})$.*

Remarks.

1. Notice that this theorem guarantees the existence of points in Q where F achieves maximum and minimum values, but it says nothing about those locations being unique. For example, there could be one or more separate points $\boldsymbol{x}_i$ where $F(\boldsymbol{x}_i) = F(\boldsymbol{x}_{\mathrm{M}})$, even though $\boldsymbol{x}_i \neq \boldsymbol{x}_{\mathrm{M}}$. The same result is of course true for the locations of the minimum values. So in short, the maximum and minimum values are unique, but their locations are not.
2. Technically, $\boldsymbol{x}_{\mathrm{m}} = \boldsymbol{x}_{\mathrm{M}}$ is possible, but then F must be constant on Q.

The previous theorem guarantees the existence of locations where minimum and maximum values occur, but it does not tell us anything about how to find these values. The next result addresses this issue, but first a little more notation is needed.

Notation. Suppose that Q is a closed, bounded region in $\mathbb{R}^n$ (so in $\mathbb{R}^2$, the region Q could be a rectangular box, a circular disk, or some other region). Then ∂Q denotes the boundary of the region (the rectangle around the box, the circle around the disk, etc.), and Q° denotes the interior of the region ($Q = Q^\circ \cup \partial Q$ since Q is closed). The interior and the boundary are shown schematically in Figure 4.3.

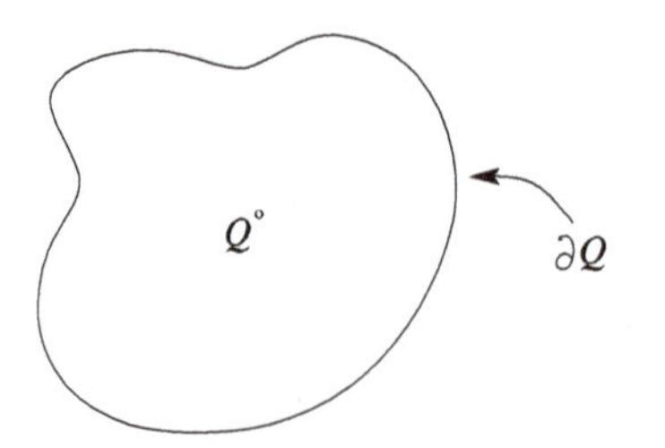

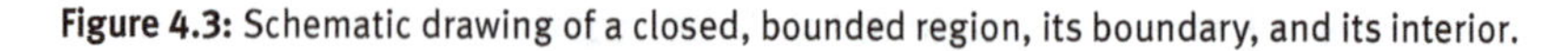

Figure 4.3: Schematic drawing of a closed, bounded region, its boundary, and its interior.

Theorem 8. *Suppose that a function $F : D \subset \mathbb{R}^n \to \mathbb{R}$ is continuous on some closed and bounded region Q of the domain D. Then $\mathbf{x}_M$ and $\mathbf{x}_m$, the locations of the maximum and minimum values, each occur in one of three distinct places:*
1. $\mathbf{x}_\bullet \in Q^\circ$ *and* $\nabla F(\mathbf{x}_\bullet) = \mathbf{0}$.
2. $\mathbf{x}_\bullet \in Q^\circ$ *and F is not differentiable at* $\mathbf{x}_\bullet$.
3. $\mathbf{x}_\bullet \in \partial Q$.

At first glance, it might seem that Theorem 8 would be difficult to prove; quite the opposite is true.

Proof. From basic topology, $Q = Q^\circ \cup \partial Q$ (i. e., a closed, bounded region is the union of its interior and its boundary), and Q° and ∂Q are disjoint. So either 3. is true, or if not, then $\mathbf{x}_\bullet \in Q^\circ$. If $\mathbf{x}_\bullet \in Q^\circ$, then either 2. is true, or if not, F is differentiable at $\mathbf{x}_\bullet \in Q^\circ$.

If F is differentiable at $\mathbf{x}_\bullet \in Q^\circ$, then $\nabla F(\mathbf{x}_\bullet)$ exists; suppose that $\nabla F(\mathbf{x}_\bullet) \neq \mathbf{0}$. Then Theorem 5, F is increasing from $\mathbf{x}_\bullet$ in the direction of $\nabla F(\mathbf{x}_\bullet)$ and F is decreasing from $\mathbf{x}_\bullet$ in the direction of $-\nabla F(\mathbf{x}_\bullet)$. Hence if $\nabla F(\mathbf{x}_\bullet) \neq \mathbf{0}$ when F is differentiable at $\mathbf{x}_\bullet$, then $\mathbf{x}_\bullet$ is neither the location of a maximum or a minimum of F. Therefore, if neither 3. or 2. are true, 1. must be true. □

Remark. All of the above works equally well if "region" is replaced by "set." But complicated sets may have complicated boundaries, so in terms of the present discussion, only regions will be covered.

Definition. A point $\mathbf{x} \in Q^\circ$ is a *critical point* for $F : \mathbb{R}^n \to \mathbb{R}$ if and only if either $\nabla F(\mathbf{x}) = \mathbf{0}$, or F is not differentiable at $\mathbf{x}$ (conditions 1. and 2. from Theorem 8 above).

Theorems 7 and 8 together show us how to find the locations of maximum and minimum values and the values themselves in many useful cases, as the next two examples demonstrate. These examples are in $\mathbb{R}^2$, but the approach presented here generalizes to $\mathbb{R}^n$. It is true, however, that if n is larger than 2 or 3, the procedure discussed here would likely require many steps.

Example 4.5. Suppose that $f(x,y) = x^3 - x^2y + y$ and suppose that $Q \subset \mathbb{R}^2$ is the triangle with corners $(0,0)$, $(3,0)$, and $(0,3)$.

Answer. In this case, f is a polynomial function, hence it is continuously differentiable on the entire plane: $f \in C^1(\mathbb{R}^2)$. So there are no critical points of type 2. from Theorem 8. If $\nabla F(\mathbf{x}_\bullet) = \mathbf{0}$, then $\partial F/\partial x(x,y) = 3x^2 - 2xy = 0$ and $\partial F/\partial y(x,y) = -x^2 + 1 = 0$. Notice that we need both of these equations to be satisfied. So from the second equation, $x = \pm 1$, and from the first equation, the two possible critical points are $(-1,-3/2)$ and $(1,3/2)$. But after consider where Q is, one sees that $(-1,-3/2) \notin Q$ so only $(1,3/2)$ is a critical point in Q.

Now we must turn our attention to ∂Q which in this case is three line segments. Notice that this leads to three single-variable maximum-minimum problems:
- $y = 0$ and $0 < x < 3$

- $x = 0$ and $0 < y < 3$
- $y = 3 - x$ and $0 < x < 3$

For the first of these segments $f(x,0) = x^3$, so $f'(x,0) = 3x^2 > 0$ on the interior of this segment, and there are no single-variable critical points on this segment. For the second, $f(0,y) = y$, so again $f'(0,y) = 1 > 0$ on this segment, and again there are no critical points. Finally, for the third, $f(x,3-x) = x^3 - x^2(3-x) + (3-x) = 2x^3 - 3x^2 - x + 3$, so $f'(x,3-x) = 6x^2 - 6x - 1 = 0$ implies that $x = 1/2 \pm \sqrt{15}/6$. Since $x = 1/2 - \sqrt{15}/6 < 0$, only $x = 1/2 + \sqrt{15}/6 \simeq 1.1455$ is in the interval $(0,3)$.

Finally, there are three more points that must be considered: the corners, that are the boundary of the boundary segments. Evaluating f at each of these points, one finds the results presented in the following table:

Locations (x,y)	$f(x,y)$
$(1,3/2)$	1
$(1.1455, 1.8545)$	0.9242
$(0,0)$	0
$(3,0)$	27
$(0,3)$	3

Theorem 8 guarantees that f achieves its maximum and minimum value at two of these five points, so these values and their location can simply be read off the list above. For the maximum, $\mathbf{x}_M = (3,0)$ and the value is $f(3,0) = 27$, while for the minimum, $\mathbf{x}_m = (0,0)$ and the value is $f(0,0) = 0$.

Remark. It may seem that the above list of possible extrema locations and function values could get rather long, but the important point is that such a list is often finite.

Example 4.6. What are the maximum and minimum values of the function $g(x,y) = 4x^2 + 9y^2 - 4x - 6y + 3$ on the circular disk of radius 1 centered at the origin?

Answer. Again, this function g is a polynomial, so a critical point can only occur where $\nabla g = \mathbf{0}$. Notice that in this case, the surface $z = g(x,y)$ is a paraboloid opening upward, so its minimum value will be at its vertex provided that this point is in the unit disk. Here, $\nabla g(x,y) = \langle 8x - 4, 18y - 6\rangle = \mathbf{0}$ imply that $(x,y) = (1/2, 1/3)$. This must be the vertex of the paraboloid, and since it is in the interior of the disk, this is the location of the minimum of g. By direct evaluation, one finds that $g(1/2, 1/3) = 1$.

The maximum must occur on the boundary of the disk, the unit circle. Because the boundary is the unit circle, polar coordinates are most helpful; here $g(\cos\theta, \sin\theta) = 4\cos^2\theta + 9\sin^2\theta - 4\cos\theta - 6\sin\theta + 3 = 5\sin^2\theta - 4\cos\theta - 6\sin\theta + 7$. On differentiating with respect to θ, one finds that $g'(\cos\theta, \sin\theta) = 10\sin\theta\cos\theta + 4\sin\theta - 6\cos\theta$. Setting this derivative equal to zero and solving this equation numerically, one sees that the two critical points on the boundary are at $\theta_1 \simeq 0.42977$ and $\theta_2 \simeq 4.4628$. Again by direct evaluation, one finds that $g(\cos\theta_1, \sin\theta_1) \simeq 1.73182$ and that $g(\cos\theta_2, \sin\theta_2) \simeq 18.497$. Therefore, the minimum value of g on this unit disk is 1 and its maximum value is 18.497.

Finally, one might wonder if the endpoints of polar coordinates (say, $\theta = 0$ and $\theta = 2\pi$) need to be checked. Because this parameterization is fully periodic, the answer is "no."

4.4 Local extrema

The previous section dealt with finding the absolute maximum and absolute minimum values for continuous functions defined on closed and bounded sets. In that case, there were often interior critical points that might have had the highest or lowest value of the function among any of the nearby points, but whose values were superseded by other points on the boundary as the absolute maximum or minimum. This section will now consider in more detail these local maxima and minima—local extrema.

Before discussing local extrema in detail, there are several new terms and bits of notation that need to be introduced. The first was briefly used earlier, but perhaps a formal definition now is in order.

Definition. Suppose that $\boldsymbol{x}_o \in \mathbb{R}^n$ is some point and $\epsilon > 0$. The *ball* of radius ϵ centered at $\boldsymbol{x}_o$ is defined as

$$B_\epsilon(\boldsymbol{x}_o) := \{\boldsymbol{x} \in \mathbb{R}^n \mid d(\boldsymbol{x}, \boldsymbol{x}_o) < \epsilon\} .$$

So all the points that are a distance less than ϵ from $\boldsymbol{x}_o$ are in this ball.

Definition. Suppose that $f : D \subset \mathbb{R}^n \to \mathbb{R}$ is a continuous function defined on some domain, D, and suppose that $\boldsymbol{x}_o \in D$. Then $\boldsymbol{x}_o$ is a *local maximum* for f if and only if there is an $\epsilon > 0$ such that $f(\boldsymbol{x}) \le f(\boldsymbol{x}_o)$ for all $\boldsymbol{x} \in B_\epsilon(\boldsymbol{x}_o)$. Also $\boldsymbol{x}_o$ is a *local minimum* for f if and only if there is an $\epsilon > 0$ such that $f(\boldsymbol{x}) \ge f(\boldsymbol{x}_o)$ for all $\boldsymbol{x} \in B_\epsilon(\boldsymbol{x}_o)$. If $\boldsymbol{x}_o$ is either a local maximum or a local minimum, it is a *local extremum*.

Notation (bra-ket notation). For n-component vectors $\boldsymbol{a} = \langle a_1, a_2, \dots a_n\rangle$ and $\boldsymbol{b} = \langle b_1, b_2, \dots b_n\rangle$, and a $n \times n$ matrix $C = [C_{ij}]$,

$$\langle \boldsymbol{a}|C|\boldsymbol{b}\rangle := \boldsymbol{a} \cdot (C\boldsymbol{b}) = \sum_{i,j=1}^{n} a_i C_{ij} b_j$$

where $C\boldsymbol{b}$ is the vector resulting from the standard matrix-vector product. This bra-ket notation is due to Dirac[2] and is widely used in quantum mechanics.

For those who are not familiar with matrix multiplication, the next example may be helpful.

---

2 Paul Dirac (1902–1984) was a theoretical physicist who work in quantum mechanics.

Example 4.7. Suppose that

$$\boldsymbol{a} = \langle 1, 2\rangle = \begin{bmatrix} 1 \\ 2 \end{bmatrix} \quad C = \begin{bmatrix} -1 & 3 \\ 2 & 7 \end{bmatrix} \quad \boldsymbol{b} = \langle b_1, b_2\rangle = \begin{bmatrix} b_1 \\ b_2 \end{bmatrix}.$$

Then

$$\langle \boldsymbol{a}|C|\boldsymbol{b}\rangle = \begin{bmatrix} 1 \\ 2 \end{bmatrix} \cdot \left(\begin{bmatrix} -1 & 3 \\ 2 & 7 \end{bmatrix} \begin{bmatrix} b_1 \\ b_2 \end{bmatrix} \right) = \begin{bmatrix} 1 \\ 2 \end{bmatrix} \cdot \begin{bmatrix} -b_1 + 3b_2 \\ 2b_1 + 7b_2 \end{bmatrix} = 3b_1 + 17b_2 .$$

So the product $C\boldsymbol{b}$ is obtained by computing the dot product of each row of C with the vector $\boldsymbol{b}$.

Definition. A $n \times n$ matrix C is *positive definite* iff for any nonzero vector $\boldsymbol{x} \in \mathbb{R}^n$, $\langle \boldsymbol{x}|C|\boldsymbol{x}\rangle > 0$. C is *negative definite* iff $\langle \boldsymbol{x}|C|\boldsymbol{x}\rangle < 0$ when $\boldsymbol{x} \neq \boldsymbol{0}$.

Example 4.8. Show that the matrix

$$\begin{bmatrix} 2 & 1 \\ 1 & 2 \end{bmatrix}$$

is positive definite.

Answer. One must show that $\langle \boldsymbol{x}|C|\boldsymbol{x}\rangle > 0$ for any $\boldsymbol{x} \neq \boldsymbol{0}$. But by direct computation,

$$\begin{aligned} \langle \boldsymbol{x}|C|\boldsymbol{x}\rangle &= \begin{bmatrix} x_1 \\ x_2 \end{bmatrix} \cdot \left(\begin{bmatrix} 2 & 1 \\ 1 & 2 \end{bmatrix} \begin{bmatrix} x_1 \\ x_2 \end{bmatrix} \right) = \begin{bmatrix} x_1 \\ x_2 \end{bmatrix} \cdot \begin{bmatrix} 2x_1 + x_2 \\ x_1 + 2x_2 \end{bmatrix} \\ &= 2x_1^2 + 2x_1x_2 + 2x_2^2 > x_1^2 + 2x_1x_2 + x_2^2 = (x_1 + x_2)^2 > 0 \end{aligned}$$

as long as $\boldsymbol{x} \neq \boldsymbol{0}$.

The key result in determining where local extrema are located is the multivariable version of the Taylor theorem.[3]

Theorem 9 (Taylor). *Suppose that a function $f : D \subset \mathbb{R}^n \to \mathbb{R}$ has continuous partial derivatives to at least second order: $f \in C^2(D)$, and suppose that $\boldsymbol{x}_o \in D$ is a base point in the domain of the function. Then if $\boldsymbol{x} \in D$ and if $\Delta\boldsymbol{x} := \boldsymbol{x} - \boldsymbol{x}_o$, one can write*

$$f(\boldsymbol{x}) = f(\boldsymbol{x}_o) + \nabla f(\boldsymbol{x}_o) \cdot \Delta\boldsymbol{x} + \frac{1}{2} \langle \Delta\boldsymbol{x}|H_f(\boldsymbol{x}_o)|\Delta\boldsymbol{x}\rangle + E(\boldsymbol{x}_o, \Delta\boldsymbol{x}) \tag{4.1}$$

where the error term $E(\boldsymbol{x}_o, \Delta\boldsymbol{x}) = o(|\Delta\boldsymbol{x}|^2)$, meaning that

$$\lim_{|\Delta\boldsymbol{x}| \to 0} \frac{E(\boldsymbol{x}_o, \Delta\boldsymbol{x})}{|\Delta\boldsymbol{x}|^2} = 0$$

and $H_f(\boldsymbol{x}_o)$ is the Hessian matrix[4] of f evaluated at $\boldsymbol{x}_o$.

---

3 Named for Brook Taylor (1685–1731) who stated but did not prove an early version of the theorem.

4 Named for German mathematician Otto Hesse (1811–1874).

One more definition is needed.

Definition. The *Hessian* matrix for $f \in C^2$ evaluated at $\boldsymbol{x}_o$ is the matrix of second partial derivatives, all evaluated at $\boldsymbol{x}_o$:

$$H_f(\boldsymbol{x}_o) := \begin{bmatrix} f_{x_1x_1}(\boldsymbol{x}_o) & f_{x_1x_2}(\boldsymbol{x}_o) & \cdots & f_{x_1x_n}(\boldsymbol{x}_o) \\ f_{x_2x_1}(\boldsymbol{x}_o) & f_{x_2x_2}(\boldsymbol{x}_o) & \cdots & f_{x_2x_n}(\boldsymbol{x}_o) \\ \vdots & \vdots & \ddots & \vdots \\ f_{x_nx_1}(\boldsymbol{x}_o) & f_{x_nx_2}(\boldsymbol{x}_o) & \cdots & f_{x_nx_n}(\boldsymbol{x}_o) \end{bmatrix}$$

Proof. The proof of this multivariable result is just an extension of the single-variable version; see, for example, Marsden and Tromba [2], Section 3.2, for a detailed proof. □

Theorem 10 (First derivative test). *Suppose that $D \subset \mathbb{R}^n$ is open and that a function $f : D \to \mathbb{R}$ has partial derivatives to at least first order on D. A point $\boldsymbol{x}_o \in D$ is the location of a local extremum of f only if $\nabla f(\boldsymbol{x}_o) = \boldsymbol{0}$, that is, $\boldsymbol{x}_o$ is a local extremum $\Rightarrow \nabla f(\boldsymbol{x}_o) = \boldsymbol{0}$.*

Proof. This result can be proven using Theorem 8, but it also follows from the Taylor theorem (Theorem 9). If $f \in C^2(D)$, the Taylor expansion (4.1) must hold using $\boldsymbol{x}_o \in D$ as the base point. If $\nabla f(\boldsymbol{x}_o) \neq \boldsymbol{0}$, then f must increase in the direction of this gradient, and must decrease in the direction opposite this gradient. □

But what if $\nabla f(\boldsymbol{x}_o) = \boldsymbol{0}$? The Taylor expansion then says that the Hessian term is the next one to consider to decide how f behaves near $\boldsymbol{x}_o \in D$, at least if $H_f(\boldsymbol{x}_o) \neq [0]$. In general, if the Hessian is positive definite, $\boldsymbol{x}_o$ is a local minimum; if the Hessian is negative definite, $\boldsymbol{x}_o$ is a local maximum; if the Hessian is neither positive definite nor negative definite, but also is nonzero, then $\boldsymbol{x}_o$ is some sort of saddle/pringle point.

Definition. Suppose that $f : D \subset \mathbb{R}^n \to \mathbb{R}$ is a continuous function defined on some domain, D, and suppose that $\boldsymbol{x}_o \in D$. Then $\boldsymbol{x}_o$ is a *saddle point* or *pringle point*[5] for f if and only if $\nabla f(\boldsymbol{x}_o) = \boldsymbol{0}$, but $\boldsymbol{x}_o$ is not a local extremum.

Theorem 11 (Second derivative test). *Suppose that $D \subset \mathbb{R}^n$ is open and that a function $f : D \to \mathbb{R}$ has continuous partial derivatives to at least second order: $f \in C^2(D)$. Suppose also that for some $\boldsymbol{x}_o \in D$, $\nabla f(\boldsymbol{x}_o) = \boldsymbol{0}$. Then if $H_f(\boldsymbol{x}_o)$ is positive definite, $\boldsymbol{x}_o$ is a local minimum; if $H_f(\boldsymbol{x}_o)$ is negative definite, $\boldsymbol{x}_o$ is a local maximum; if $H_f(\boldsymbol{x}_o)$ has both positive and negative eigenvalues, then $\boldsymbol{x}_o$ is a saddle point.*

Proof. This result follows directly from the Taylor theorem, Theorem 9. The details are a bit beyond the scope of our discussion, but can be found in Marsden and Tromba [2], Section 3.3, at least when $f \in C^3(D)$. □

5 So named because the corresponding surface near $\boldsymbol{x}_o$ often looks like a Pringles' potato crisp.

The second derivative test has a simpler form, at least in two dimensions when $f : D \subset \mathbb{R}^2 \to \mathbb{R}$.

Corollary 3 (Second derivative test for $D \subset \mathbb{R}^2$). *Suppose that $D \subset \mathbb{R}^2$ is open and that a function $f : D \to \mathbb{R}$ has continuous partial derivatives to at least second order: $f \in C^2(D)$. Suppose also that for some $\mathbf{x}_o \in D$, $\nabla f(\mathbf{x}_o) = \mathbf{0}$. Let $\mathcal{D} := \det(H_f(\mathbf{x}_o)) \equiv (f_{xx}(\mathbf{x}_o))(f_{yy})(\mathbf{x}_o) - (f_{xy}(\mathbf{x}_o))^2$ be the* discriminant *(the determinant of the Hessian).*
- *If $D > 0$ and $f_{xx}(\mathbf{x}_o) > 0$, then $\mathbf{x}_o$ is a local minimum.*
- *If $D > 0$ and $f_{xx}(\mathbf{x}_o) < 0$, then $\mathbf{x}_o$ is a local maximum.*
- *If $D < 0$, then $\mathbf{x}_o$ is a saddle point.*
- *If $D = 0$, then the local behavior is determined by higher order terms.*

Let us now look at a number of examples:

Example 4.9. Suppose that $f(x, y) = x^2+3y^2-2x-12y+13$. Please find and characterize the critical points of f.

Answer. Since this function is a polynomial, it is everywhere differentiable, so the only critical points occur where $\nabla f(\mathbf{x}) = \mathbf{0}$. Here $\nabla f(\mathbf{x}) = \langle 2x-2, 6y-12\rangle$, thus the only critical point is $\mathbf{x}_o = (1, 2)$. From the second derivative test, $\forall (x, y) \in \mathbb{R}^2$,

$$\det(H_f)(1,2) = \begin{vmatrix} 2 & 0 \\ 0 & 6 \end{vmatrix} = 12 > 0$$

Thus $(1, 2)$ is a local minimum. Of course, for this polynomial function, one could also have arrived at this result by completing the square in x and y separately and finding that the surface $z = f(x, y)$ is an elliptical paraboloid that opens upward.

Example 4.10. Suppose that $f(x, y) = 2x^3 - y^2 - 6x - 2y$. Please find and characterize the critical points of f.

Answer. Again since this function is a polynomial, it is everywhere differentiable, so the only critical points occur where $\nabla f(\mathbf{x}) = \mathbf{0}$. Now $\nabla f(\mathbf{x}) = \langle 6x^2 - 6, 2y - 2\rangle$, so critical points occur at $(-1, 1)$ and $(1, 1)$. In this case,

$$\det(H_f)(-1,1) = \begin{vmatrix} 12(-1) & 0 \\ 0 & 2 \end{vmatrix} = -24 < 0$$

$$\det(H_f)(1,1) = \begin{vmatrix} 12(1) & 0 \\ 0 & 2 \end{vmatrix} = 24 > 0$$

The first of these calculations implies that $(-1, 1)$ is a saddle (pringle) point. Because both $f_{xx}(1, 1) > 0$ and $f_{yy}(1, 1) > 0$, the second implies that $(1, 1)$ is a local minimum.

4.5 Lagrange multipliers

There is an interesting connection between gradients and extrema; it was discovered by Joseph-Louis Lagrange[6] and bears his name: Lagrange multipliers. The method of Lagrange multipliers allows one to efficiently find extreme values for functions subject to constraints (side conditions). A simple example will help illustrate this method.

Example 4.11. What is the minimum value of $f(x, y) = x^2 + y^2$ on the parabola $g(x, y) = x^2 + 3x + 3 - y = 0$?

Answer. Based on what we have discussed so far, one might solve this problem by solving $g(x, y) = 0$ for y, plugging this expression for y into $f(x, y)$, and then differentiating the resulting expression in x with respect to x and setting this derivative equal to zero. Following this approach will work, and the result is that the minimum occurs at $(-1, 1)$. (This must be a minimum because f is unboundedly positive on this parabola.) But notice what also happens at this point; consider Figure 4.4. The gradient of the function f is a scalar multiple of the gradient of the constraint function g. Suppose that c_m is the value of f at the point where these gradients line up. The level curves of f for $c > c_m$ cross the constraint curve transversely; those for $c < c_m$ miss the con-

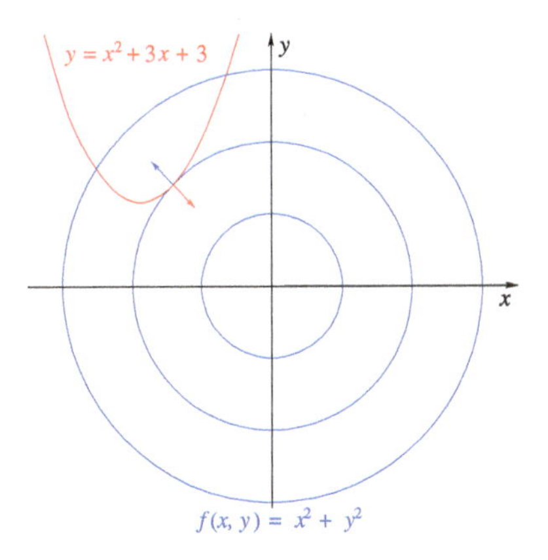

Figure 4.4: An example of how Lagrange multipliers works. The red curve is the constraint $g(x, y) = 0$ ($y = x^2 + 3x + 3$); the blue curves are the level curves of $f(x, y) = x^2 + y^2$. The red and blue vectors are the gradients for the g and f, respectively. When the constraint curve and a level curve are tangent (and, therefore, when their gradients are parallel), f achieves its minimum value on the constraint curve.

6 Joseph-Louis Lagrange (1736–1813) was a French/Italian mathematician who worked for 20 years in Berlin. Along with defining Lagrange multipliers, Lagrange contributed to several areas of mathematics, particularly calculus of variations and mechanics.

straint curve altogether. Only when f achieves its extreme value are the curves tangent and the gradients parallel.

What Lagrange realized is the following theorem:

Theorem 12 (Lagrange). *Suppose that for some domain $D \subset \mathbb{R}^2$, the functions $f, g : D \to \mathbb{R}$, and suppose that both f and g are continuously differentiable on their domain: $f, g \in C^1(D)$. Suppose also that c is in the range of g, and consider the constraint curve C defined by $g(x,y) = c$. Finally, suppose that $\nabla g(x_o, y_o) \neq 0$ for some point $(x_o, y_o) \in C$. If f restricted to C has an extreme value at (x_o, y_o), then*

$$\nabla f(x_o, y_o) = \lambda \nabla g(x_o, y_o)$$

for some $\lambda \in \mathbb{R}$. So extrema occur only when these gradients are parallel.

Remark. The real scalar λ in Theorem 12 is the *Lagrange multiplier*. Notice that the Lagrange multiplier might be zero.

Example 4.11. What is the minimum value of $f(x,y) = x^2 + y^2$ on the parabola $g(x,y) = x^2 + 3x + 3 - y = 0$?

Answer. Now let us use Lagrange multipliers to answer this question: Here, $\nabla f(x,y) = 2\langle x, y\rangle$ and $\nabla g(x,y) = \langle 2x+3, -1\rangle$. So the minimum of f must occur when

$$\begin{aligned} 2x &= \lambda(2x+3) \\ 2y &= \lambda(-1) \\ y &= x^2 + 3x + 3 \end{aligned}$$

(the first two equations above come from the gradients being parallel; the third is the constraint). Solving these three equations simultaneously, one finds that $(x,y) = (-1, 1)$ and that $\lambda = -2$. Notice that the exact value of the Lagrange multiplier is not really important in answering the question.

The Lagrange multiplier method can be used in higher dimensions; the theorem for $n = 3$ is similar to Theorem 12 above, except that the constraint is now a surface rather than a curve.

Theorem 13 (Lagrange). *Suppose that for some domain $D \subset \mathbb{R}^3$, the functions $f, g : D \to \mathbb{R}$, and suppose that both f and g are continuously differentiable on their domain: $f, g \in C^1(D)$. Suppose also that c is in the range of g, and consider the constraint surface C defined by $g(x,y,z) = c$. Finally, suppose that $\nabla g(x_o, y_o, z_o) \neq 0$ for some point $(x_o, y_o, z_o) \in C$. If f restricted to C has an extreme value at (x_o, y_o, z_o), then*

$$\nabla f(x_o, y_o, z_o) = \lambda \nabla g(x_o, y_o, z_o)$$

for some $\lambda \in \mathbb{R}$. So extrema occur only when these gradients are parallel.

Proof. The proof of this result is a variation on the proof of Theorem 6; see Exercise 4.17. □

Example 4.12. What is the minimum value of $f(x,y,z) = x^2 + 2y^2 - z$ on the plane $z = 2x + 3y$?

Answer. To use Lagrange multipliers (or the Lagrange multiplier method), one first needs to write the constraint in the form $g(x,y,z) = 0$. In this case, the constraint can be $g(x,y,z) = 2x + 3y - z = 0$. So $\nabla f(x,y,z) = \langle 2x, 4y, -1\rangle$ and $\nabla g(x,y,z) = \langle 2,3,-1\rangle$; the Lagrange multiplier system that must be solved is

$$\langle 2x, 4y, -1\rangle = \lambda\langle 2,3,-1\rangle\,, \qquad z = 2x + 3y$$

In this case, the system is easy to solve: $\lambda = 1$, $x = 1$, $y = 3/4$ and $z = 17/4$. The minimum value of f is then $f(1, 3/4, 17/4) = -17/8$. This value is certainly a minimum (rather than a maximum) since $f(0,0,0) = 0$.

For this particular example, one could solve the problem without using Lagrange multipliers by using the constraint to eliminate z and then completing the squares:

$$f(x,y,2x+3y) = x^2 + 2y^2 - 2x - 3y = (x-1)^2 + 2(y-3/4)^2 - 17/8$$

Exercises 4

4.1. For each of the following functions F, please describe the level curves or level surfaces corresponding to $F(x,y) = c$ or $F(x,y,z) = c$, and please state which values of c correspond to curves or surfaces.

(a) $F(x,y) = 1 - x^2 - y^2$
(b) $F(x,y) = x^3 - 4y + 3$
(c) $F(x,y) = mx + b - y$ where m, b are constants
(d) $F(x,y) = \sin xy$
(e) $F(x,y) = x^2 - 4x + y - 8$
(f) $F(x,y,z) = 5x - 3y + 2z$
(g) $F(x,y,z) = x^2 + 2y^2 + 3z^2$
(h) $F(x,y,z) = \ln(x^2 + 7y^2 + 1) - z$

Answer. (c) For any value of c, a family of lines with slope m and y-intercept $b - c$. (g) A family of ellipsoids centered at $(0,0,0)$ provided that $c > 0$.

4.2. What can be said about the level hypersurfaces for the function defined by $F(x_1,x_2,x_3,x_4) = x_1^2 + x_2^2 - x_3^2 + x_4$?

4.3. Find an equation for the plane tangent to the surface $F(x,y,z) = 5xy^2z^2 - x^2y^3 + 3x^3yz = 25$ at the point $(x,y,z) = (1,1,2)$.

Answer. $36(x-1) + 43(y-1) + 23(z-2) = 36x + 43y + 23z - 125 = 0$

4.4. For $F(x,y,z) = 2y^3 + 3x^2y - xyz + 5yz^2 - 9$, please find the equation of the plane tangent to the surface $F(x,y,z) = 0$ at $(-1,1,-1)$.

4.5. For the ellipsoid $x^2 + 9y^2 + 4z^2 = 36$, please find the equations of the two tangent planes whose normal vector is $\mathbf{N} = \langle 1, 3, 2 \rangle$.

4.6. Please find an equation for the plane tangent to the surface $F(x, y, z) = xz - x^2y + yz^2 = 1$ at the point $(x, y, z) = (x_o, y_o, z_o)$ assuming that this point is on the surface.

4.7. Suppose that $f : \mathbb{R}^2 \to \mathbb{R}$ is continuously differentiable on its domain and that $f(x_o, y_o) = 0$. Please show that $\nabla f(x_o, y_o)$ is perpendicular to the level curve $f(x, y) = 0$ provided that this gradient is nonzero. **Hint:** Parameterize the curve by the vector function $\mathbf{x}(t) = \langle x_1(t), x_2(t) \rangle$ and use a version of the chain rule argument used to show that ∇F is perpendicular to the surface $F(x, y, z) = 0$.

4.8. Find the maximum and minimum values and their location for following functions f:

(a) $f(x, y) = x^2 + 9y^2 - 2x - 12y + 8$ on the closed triangular domain whose corners are $(0, 0)$, $(4, 0)$, and $(0, 5)$.

(b) $f(x, y) = x^2 + 9y^2 - 2x - 12y + 8$ on the closed circular disk of radius 5 centered at the origin.

(c) $f(x, y) = (1 - x + y)/(1 + x^2 + y^2)$ on the closed circular disk of radius 2 centered at the origin. **Hint:** Define $g(r, \theta) := f(r\cos\theta, r\sin\theta)$ and solve the entire problem in polar coordinates.

4.9. What points on the circle $(x-2)^2+(y-3)^2 = 1$ and on the line $y = -x-1$ are closest to each other? (Which point on this circle is closest to which point on this line?) **Hint:** To use the concepts discussed in this chapter to solve this problem, let (x, y) be a point on the circle and (X, Y) be a point on the line, and minimize the square of the distance between these points. As an alternative, can you think of a geometric way to solve this problem?

4.10. What points on the ellipse $\frac{(x-2)^2}{4} + (y - 5)^2 = 1$ and on the line $y = -2x - 3$ are closest to each other? (Which point on this ellipse is closest to which point on this line?) **Hint:** As in the previous exercise, let (x, y) be a point on the ellipse and (X, Y) be a point on the line, and minimize the square of the distance between these points. Again, can you think of a geometric way to solve this problem? Notice that geometry can be helpful in simplifying calculations.

4.11. Please compute the Hessian matrix for $F(x, y, z) = 2x^2 - xy + 3y^2z + 3xz - z^3$.

4.12. Please find and classify the local extrema for the following functions:

(a) $f(x, y) = 3x^2 + y^2 - 2xy$

(b) $f(x, y) = 5x^2 + 3y^2 - 2xy + 2x - 6y$

(c) $f(x, y) = 3x^2 + 12xy + y^2 - 2x + 3y$

(d) $f(x, y) = e^{-x^2-2x-y^2}$

(e) $f(x, y) = \frac{1-x+y}{1+x^2+y^2}$

Answer. (a) Local and global minimum at $(0,0)$ where $f(0,0) = 0$; no maximum. (d) No minimum; local and global maximum at $(-1,0)$ where $f(-1,0) = e$.

4.13. Suppose that

$$f(x,y) = \frac{2}{\sqrt{4 - x^2 - y^2}}.$$

What is the domain for f? Does f achieve a maximum or minimum on its domain? How does this relate to Theorem 7? Explain your thinking.

4.14. What is the maximum value of $f(x,y) = -x^2 - y^2$ on the parabola $g(x,y) = x^2 - 4x + 3 + y = 0$? Where does this maximum occur?

4.15. What is the maximum value of $F(x,y,z) = 2x+4y+z$ on the paraboloid $G(x,y,z) = x^2 + y^2 + z = 0$?

Answer. $F(1,2,5) = 4$

4.16. Please find the minimum distance between the point $(1,2,3)$ and the hyperboloid $x^2 + y^2 - z^2 = 1$.

4.17. Please prove the Lagrange multiplier theorem (Theorem 13). **Hint:** Suppose that the vector function $\mathbf{x}(t)$ traces out a curve C which lies on the surface $g(x,y,z) = 0$ and that $\mathbf{x}(0) = \langle x_0, y_0, z_0 \rangle$. Now differentiate $g(\mathbf{x}(t)) = 0$ with respect to t to show that (as in Theorem 6) $\nabla g(x_0, y_0, z_0)$ is perpendicular to the surface. Now explain why $\nabla f(x_0, y_0, z_0)$ must point in the same (or opposite) direction to $\nabla g(x_0, y_0, z_0)$ by showing that both are perpendicular to $\dot{\mathbf{x}}(0)$.

4.18. Solve Exercise 4.9 using Lagrange multipliers.

4.19. Solve Exercise 4.10 using Lagrange multipliers.

5 Multiple integrals-integration in $\mathbb{R}^n$

5.1 Riemann integration versus iterated integrals

5.1.1 Single-variable Riemann integration

In principle, the definition for Riemann[1] integration in $\mathbb{R}^n$ is essentially the same as that of Riemann integration in $\mathbb{R}^1$, so let us begin by reviewing the classic definition of the Riemann integral from single-variable calculus. Suppose that $f : [a,b] \to \mathbb{R}$, that is, f is a real-valued function defined on an interval $[a,b] \subset \mathbb{R}$. Suppose further that f is *bounded* on $[a,b]$, meaning that there is a fixed real number $M > 0$ such that $|f(x)| < M$ for all $x \in [a,b]$. Let P be a *partition* of this interval: $P := \{x_0, x_1, x_2, \ldots, x_n\}$ where $x_0 = a, x_n = b$ and for all i, $x_{i-1} < x_i$. So P divides the interval into n subintervals. Now define $\Delta x_i := x_i - x_{i-1}$ for all i, and define the *norm* of the partition P as the length of the longest subinterval: $|P| := \max_i \Delta x_i$. Finally, let SP be a set of sampling points for this partition: $SP := \{\xi_1, \xi_2, \ldots, \xi_n\}$ where for all i, $\xi_i \in [x_{i-1}, x_i]$. So each subinterval of the partition has a corresponding sampling point in SP. The function is sampled at each of these points, and the value $f(\xi_i)$ is used as the representative value of the function for the entire subinterval. For a given partition P and a given sampling-point set SP, the *Riemann sum* $R(f,P,SP)$ is

$$R(f,P,SP) := \sum_{i=1}^{n} f(\xi_i)\Delta x_i$$

which is shown in Figure 5.1.

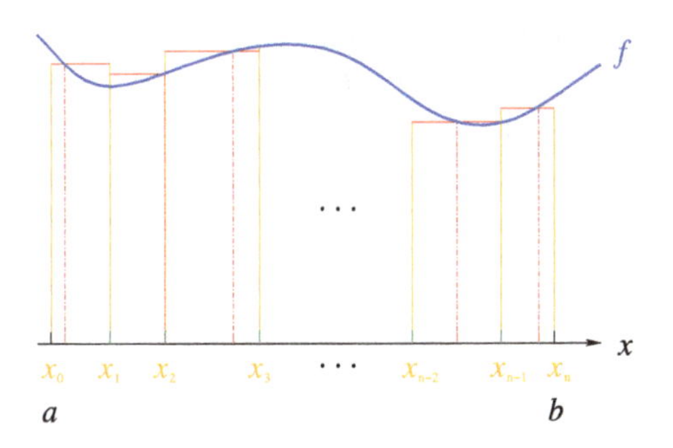

Figure 5.1: A Riemann sum for a continuous function f where $f > 0$. The partition $P := \{a = x_0, x_1, x_2, \ldots, x_n = b\}$ is indicated by the green vertical segments, and the sampling points are indicated by the red vertical segments. Notice that for the second subinterval, the sampling point is at the right endpoint of that subinterval. Where on the diagram are the sampling points ξ_i for each of the other subintervals? The value of the Riemann sum is the total area of the red/yellow rectangles.

1 Named in honor of Bernhard Riemann (1826–1866), a German mathematician who first carried out much of the rigorous work on this type of integration in the middle of the 19th century.

https://doi.org/10.1515/9783110660609-005

Definition. The *Riemann integral* of the function f over the interval $[a, b]$ is

$$\int_a^b f(x)\,dx := \lim_{|P|\to 0} R(f, P, SP) = \lim_{|P|\to 0} \sum_{i=1}^{n} f(\xi_i)\Delta x_i$$

provided that this limit exits no matter how the partition is chosen and no matter how the sampling points are chosen, as long as the partition is sufficiently fine (i. e., its norm is sufficiently small). Any function f for which this limit exists and is finite for a given interval $[a, b]$ is said to be *Riemann integrable* on $[a, b]$.

Assuming that a certain function f is Riemann integrable on $[a, b]$, that is, that the above limit converges to a finite value for all partitions and all sampling sets, then one can compute it directly by choosing a uniform partition with $\Delta x = (b-a)/n$ being the width of each subinterval, and by sampling each subinterval using the right endpoint of each subinterval: $\xi_i = x_i$. As it turns out, continuous functions are always Riemann integrable.

Example 5.1. Using the definition of the Riemann integrable with a uniform partition, please compute the value of

$$\int_1^3 x^2\,dx\,.$$

Answer. In this case, $a = 1$, $b = 3$, $f(x) = x^2$, $\Delta x = (b-a)/n = 2/n$, and $x_i = a + i\Delta x = 1 + 2i/n$. Using the right subinterval endpoint as the sampling point, one has that $\xi_i = x_i = 1 + 2i/n$. Because the partition is uniform, $|P| = (b-a)/n = 2/n$, and in this uniform case, $|P| \to 0$ is equivalent to $n \to \infty$. Hence

$$\begin{aligned}
\int_1^3 x^2\,dx &= \lim_{|P|\to 0} \sum_{i=1}^{n} \left(1 + i\frac{2}{n}\right)^2 \frac{2}{n} \\
&= \lim_{n\to\infty} \frac{2}{n} \sum_{i=1}^{n} \left(1 + i\frac{4}{n} + i^2\frac{4}{n^2}\right) \\
&= \lim_{n\to\infty} \frac{2}{n} \left(\sum_{i=1}^{n} 1 + \frac{4}{n}\sum_{i=1}^{n} i + \frac{4}{n^2}\sum_{i=1}^{n} i^2\right) \\
&= \lim_{n\to\infty} \frac{2}{n} \left(n + \frac{4}{n}\frac{n(n+1)}{2} + \frac{4}{n^2}\frac{n(n+1)(2n+1)}{6}\right) \\
&= \lim_{n\to\infty} \left(2 + 4\left(1 + \frac{1}{n}\right) + \frac{4}{3}\left(1 + \frac{1}{n}\right)\left(2 + \frac{1}{n}\right)\right) \\
&= 2 + 4 + \frac{8}{3} = \frac{26}{3}\,.
\end{aligned}$$

The above calculation uses two standard summation formulas[2] which can be proven by induction:

- $\sum_{i=1}^{n} i = \frac{n(n+1)}{2}$
- $\sum_{i=1}^{n} i^2 = \frac{n(n+1)(2n+1)}{6}$

Of course, anyone who has studied calculus knows that the long computation in the previous example is not necessary to evaluate the given integral. By the fundamental theorem of calculus, the integral is just

$$\int_1^3 x^2\,dx = \left(\frac{x^3}{3}\right|_1^3 = 9 - \frac{1}{3} = \frac{26}{3}.$$

This indicates something that will be demonstrated many times across this chapter and the rest of this book: The definition of the Riemann integral is important to establish many of its basic properties and to understand what it represents, but it is seldom used to evaluate an integral. One almost always uses powerful theorems to evaluate integrals.

5.1.2 Multivariable Riemann integration

The generalization of the definition of the Riemann integral from $\mathbb{R}^1$ to $\mathbb{R}^n$, at least when the domain of integration is rectangular, follows immediately from the single-variable definition. For simplicity, assume that $n = 2$. Let $R := [a,b] \times [c,d]$ be a rectangle in $\mathbb{R}^2$. Suppose that $f : R \to \mathbb{R}$, that is, f is a real-valued function defined on the rectangle R. Suppose further that f is bounded on R. Let P_1 be a partition of $[a,b]$, that is, $P_1 := \{x_o, x_1, x_2, \ldots, x_n\}$ where $x_o = a$, $x_n = b$, and for all i, $x_{i-1} < x_i$, and let P_2 be a partition of $[c,d]$, that is, $P_2 := \{y_o, y_1, y_2, \ldots, y_m\}$ where $y_o = c$, $y_m = d$ and for all j, $y_{j-1} < y_j$. So $P_1 \times P_2$ divides the rectangle into nm subrectangles. Again define $\Delta x_i := x_i - x_{i-1}$ for all i, and in addition, define $\Delta y_j := y_j - y_{j-1}$ for all j. Define $\Delta A_{ij} := \Delta x_i \Delta y_j$. Let $P := P_1 \times P_2$ be a partion of the rectangle, and define the *norm* of P as the largest subrectangle length or width: $|P| := \max_{ij}\{\Delta x_i, \Delta y_j\}$. Finally, let SP be a set of sampling points for this partition: $SP := \{\xi_{11}, \xi_{21}, \xi_{12}, \ldots, \xi_{nm}\}$ where for all (i,j), $\xi_{ij} \in [x_{i-1}, x_i] \times [y_{j-1}, y_j]$. Thus again each subrectangle has a corresponding sampling point.

Definition. The *Riemann integral* of the function f over the rectangle R is

$$\iint_R f(x,y)\,dA := \lim_{|P|\to 0} \sum_{i=1}^{n} \sum_{j=1}^{m} f(\xi_{ij}) \Delta A_{ij}$$

provided this limit exits no matter how the partitions are chosen and no matter how the sampling points are chosen, as long as the partition is sufficiently fine (i. e., its

2 The first formula has been known for hundreds of years, but according to legend, as a child Gauss rediscovered it, and used it to avoid the punishment of having to add up the first hundred integers.

norm is sufficiently small). Any function f for which this limit exists and is finite for a given rectangle R is said to be *Riemann integrable* on R.

Remarks.

1. For integration over rectangular prisms (boxes) in $\mathbb{R}^n$ for $n \geq 3$, the definition is extended in the obvious way: the prism is partitioned into subprisms with sufficiently small edge lengths, and the integrand f is sampled at points inside each subprism. If the sum of the products of these sampled function values and the volumes of each subprism approaches a fixed, finite value as the edge length decreases, no matter how the partitioning or sampling is done, then this f is Riemann integrable over the given rectangular prism, and this fixed, finite value is the value of the integral.
2. As was mentioned above, this definition is *not* very helpful in evaluating integrals. Its main use is in proving basic results about integrals. In particular, one can use this definition to show that if f is continuous on a given rectangle, then f is integrable. *Most* of the functions we consider in this chapter are continuous, and thus integrable.
3. Strictly speaking, the definition given above only allows integration over rectangles, and by extension, rectangular prisms; frequently, we will need to integrate over more general domains in $\mathbb{R}^2$ and $\mathbb{R}^3$. Although the details are somewhat technical and beyond the scope of this text, our definition can be extended to domains that can be well approximated by rectangles or rectangular prisms. Riemann integrals over domains in $\mathbb{R}^2$ are called *double integrals*; those over domains in $\mathbb{R}^3$ are called *triple integrals*.

5.1.3 Iterated integrals

In addition to extending the definition of the single-variable Riemann integral to domains in $\mathbb{R}^2$, $\mathbb{R}^3$ or in general $\mathbb{R}^n$ in the manner discussed above, there is another type of extension—iteration. Here, one simply computes single-variable integrals sequentially, moving from the inner-most integral outward. So for example, by definition,

$$\int_a^b \int_c^d f(x,y)\,dy\,dx := \int_a^b \left(\int_c^d f(x,y)\,dy \right) dx$$

where for the inner integral, the outer variable x is treated as a constant. A specific example is useful to clarify this definition.

Example 5.2. Suppose that $f(x,y) = x^2(y+1)$. Please evaluate

$$\int_0^1 \int_3^5 f(x,y)\,dy\,dx\,.$$

Answer. This is an iterated integral, so by definition,

$$\int_0^1\int_3^5 f(x,y)\,dy\,dx = \int_0^1\left(\int_3^5 x^2(y+1)\,dy\right)dx$$
$$= \int_0^1\left(x^2\int_3^5 y+1\,dy\right)dx$$
$$= 10\int_0^1 x^2\,dx = \frac{10}{3}.$$

Notice that iterated integrals never involve higher-dimensional Riemann sums. They are simply several single-variable integrals evaluated one after the other.

5.1.4 The Fubini theorem and the relationship between Riemann and iterated integrals

The examples above suggest the following question: "When do iterated integrals agree with the corresponding Riemann integrals?" Notice that when these integrals agree, we can evaluate the Riemann integral by evaluating a corresponding iterated integral as we did in Example 5.2. The most general answer to this question is in fact complicated and hinges on more advanced concepts from measure theory. But when the integrand f is continuous on the domain of integration R, a famous result known as the Fubini theorem gives the needed answer.

Theorem 14 (Fubini[3]). *If f is continuous on a rectangle $R := [a,b]\times[c,d]\subset\mathbb{R}^2$, then*

$$\iint_R f(x,y)\,dA = \int_a^b\left(\int_c^d f(x,y)\,dy\right)dx = \int_c^d\left(\int_a^b f(x,y)\,dx\right)dy.$$

Here is a simple example that uses the Fubini theorem:

Example 5.3. For $R = [1,3]\times[2,5]$, please compute

$$\iint_R x^2y + xy^2\,dA.$$

Answer. By using the Fubini theorem, this double integral can be computed as two single integrals in either order. When integrating with respect to one variable, all other

3 Named in honor of Italian mathematician Guido Fubini (1879–1943). The result that Fubini actually proved involves measure theory and Lebesgue integration. This result was known before Fubini's work, but nonetheless, has come to be known by his name.

variables are treated as constants. Thus

$$\iint_R x^2y + xy^2\, dA = \int_1^3 \left(\int_2^5 x^2y + xy^2\, dy \right) dx = \int_1^3 \left(\frac{x^2y^2}{2} + \frac{xy^3}{3} \right|_{y=2}^{y=5} dx$$
$$= \int_1^3 \frac{21}{2}x^2 + 39x\, dx = \left(\frac{7}{2}x^3 + \frac{39}{2}x^2 \right|_{x=1}^{x=3} = 247\,.$$

In the next example, it is important to realize that the x-integral should be computed first, even though Fubini's theorem guarantees that the value is the same in either order.

Example 5.4. For $R := [\pi/2, 3\pi/2] \times [-2, 1]$, please compute

$$\iint_R y \sin xy\, dA\,.$$

Answer. Because the integrand for this integral is continuous on the domain of integration R, the Fubini theorem can be applied to evaluate this integral:

$$\iint_R y \sin xy\, dA = \int_{-2}^{1} \left(\int_{\frac{\pi}{2}}^{\frac{3\pi}{2}} y \sin(xy)\, dx \right) dy$$
$$= \int_{-2}^{1} \left(\int_{\frac{\pi}{2}y}^{\frac{3\pi}{2}y} \sin u\, du \right) dy$$
$$= \int_{-2}^{1} \left(-\cos u \right|_{\frac{\pi}{2}y}^{\frac{3\pi}{2}y} dy$$
$$= -\int_{-2}^{1} \cos\left(\frac{3\pi}{2}y\right) - \cos\left(\frac{\pi}{2}y\right) dy$$
$$= -\frac{2}{3\pi} \sin\left(\frac{3\pi}{2}y\right|_{-2}^{1} + \frac{2}{\pi} \sin\left(\frac{\pi}{2}y\right|_{-2}^{1}$$
$$= \frac{8}{3\pi}$$

Evaluation of the inner integral above uses the substitution $u = xy$.

Remark. Notice that the Fubini theorem says that the result will be the same regardless of whether the x-integral or the y-integral is computed first (is on the inside). In this example, however, because of the nature of the integrand, it is much easier to compute the x-integral first. If one computes the y-integral first, one quickly faces the integrand $u \sin u$ which requires integration by parts to compute by hand. Indeed it is possible that picking the wrong order can lead to an integral that can not be computed by hand, while picking the right order yields an integral that is relatively easy to compute. If the initial order does not work, consider reversing the order.

5.1.5 When it all goes wrong: functions that are not Riemann integrable

Which functions are Riemann integrable? Perhaps surprisingly, this question is difficult to answer exactly. The limit in the definition of the Riemann integral exists for continuous functions, piecewise continuous functions and many other important functions, but by no means for all functions. Even proving that continuous functions are Riemann integrable requires relatively advanced concepts; a rigorous proof did not exist until the concept of compactness was understood in the second half of the 19th century. To study the details of why continuous functions are Riemann integrable, one can see, for example, Rosenlicht [4], Rudin [5], or Kosmala [1].

While continuity everywhere on the integration domain is not necessary for a function to be Riemann integrable, a complete lack of continuity is a serious problem. There is, for example, one easily-defined function where the limit in the definition of the Riemann integral does not exist.

Example 5.5. Consider the characteristic function for the rationals:

$$\mathcal{X}_{\mathbb{Q}}(x) := \begin{cases} 1 & x \in \mathbb{Q} \\ 0 & x \notin \mathbb{Q} \end{cases}$$

In words, $\mathcal{X}_{\mathbb{Q}}(x)$ is 1 when x is rational, and 0 when x is irrational. What happens when the limit in the definition of the Riemann integral is applied to this characteristic function over the interval $[0,1]$? Consider any partition P of $[0,1]$. Notice that the value of the sum over P depends not on the norm of the partition, but rather whether the sampling is done at rational or irrational numbers (since there are both rational and irrational numbers in each subinterval of any partition). That is, if one samples only at irrational numbers, then $\xi_i \notin \mathbb{Q}$ for all i, $1 \le i \le n$, and $\mathcal{X}_{\mathbb{Q}}(\xi_i) = 0$, implying that the sum is zero. If, on the other hand, one samples only at rational numbers, then $\xi_i \in \mathbb{Q}$ for all i, $1 \le i \le n$, and $\mathcal{X}_{\mathbb{Q}}(\xi_i) = 1$. In the first case, every Riemann sum must be zero no matter what partition is used, and in the second case, every Riemann sum must be one. Thus the limit cannot exist, and this function is *not* Riemann integrable.

In many ways, the problem with the integration in the previous example is with the definition of the Riemann integral, not with the function $\mathcal{X}_{\mathbb{Q}}$. Years after the work of Riemann, it was understood that a more sophisticated definition for the integral was possible, and under this latter definition, the value of the integral of $\mathcal{X}_{\mathbb{Q}}$ over any interval is zero. Discussion of this newer integral, the Lebesgue[4] integral, is beyond the scope of this book. Nonetheless, the defining of the Lebesgue integral was a major step forward in our understanding of this area of mathematics. Indeed Lebesgue measure also gives the amount of continuity that a function must possess to be Riemann integrable: a function is Riemann integrable if and only if its set of discontinuities has Lebesgue measure zero.

4 Henri Lebesgue (1875–1941) was a French mathematician famous for extending the concept of integration in his 1902 PhD thesis.

5.2 Double integrals: integration over domains in $\mathbb{R}^2$

In the previous section, iterated integrals where used to evaluate Riemann integrals over rectangles in $\mathbb{R}^2$. Now iterated integrals will be used to evaluate Riemann integrals over more general domains in $\mathbb{R}^2$. As the Fubini theorem was first stated, it would apply only to continuous functions and only to rectangular domains. Neither of these two assumptions, however, is essential, and we will now use the Fubini theorem for more complicated domains and for piecewise continuous integrands.

5.2.1 Integration using rectangular coordinates

Consider the double integral

$$\iint_D f(x,y)\,dA$$

where the integration domain D is bounded by two piecewise smooth curves given by the expressions $y = c(x)$ and $y = d(x)$ over some interval $[a,b]$ where $c(x) \leq d(x)$ and c and d are continuous and piecewise differentiable functions (see Figure 5.2). So in this case, $D = \{(x,y) \mid a \leq x \leq b,\ c(x) \leq y \leq d(x)\}$, and our double integral can be evaluated as an iterated integral.

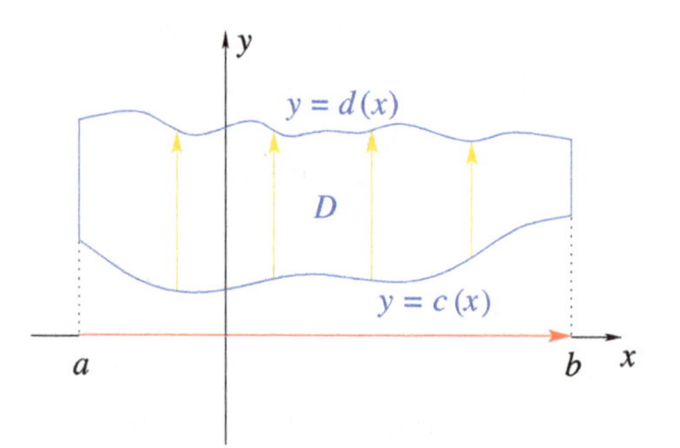

Figure 5.2: Domain of integration D between two curves, $y = c(x)$ and $y = d(x)$, with $c(x) \leq d(x)$ and $a \leq x \leq b$. For this domain, the y-integral should be inside, and the x-integral should be outside. The gold rays indicate how the y-integration depends on x: for each x-value corresponding to D, the ray begins at $y = c(x)$ and ends at $y = d(x)$. The single red ray indicates the final x-integration.

Theorem 15 (Fubini). *If f is continuous on the domain D described above, then*

$$\iint_D f(x,y)\,dA = \int_a^b \left(\int_{c(x)}^{d(x)} f(x,y)\,dy \right) dx\,.$$

Remarks.

1. Of course if $D = \{(x,y) \mid c \le y \le d,\ a(y) \le x \le b(y)\}$ for continuous and piecewise differentiable functions a and b defined for $y \in [c,d]$, then
$$\iint\limits_D f(x,y)\,dA = \int_c^d \left(\int_{a(y)}^{b(y)} f(x,y)\,dx \right) dy\,.$$
And if our domain D can be describe in both of these ways, then one has a choice of how to set up the iterated integrals.
2. At times, it will be useful to write an iterated integral without giving the specific integration limits for each variable, but rather just stating the domain as is done for a double integral. In this case, the differentials at the end of the integral will indicate the type of integral; dA still indicates a double integral, while $dy\,dx$ or $dx\,dy$ indicates an iterated integral. Thus the Fubini theorem can be written as
$$\iint\limits_D f(x,y)\,dA = \iint\limits_D f(x,y)\,dy\,dx = \iint\limits_D f(x,y)\,dx\,dy$$
provided that D is bounded and its boundary is rectifiable so that we can make sense of the iterated integrals.

Example 5.6. For the triangular integration domain bounded by the lines $y = x$, $y = 0$, and $x = 1$ (see Figure 5.3), please compute
$$\iint\limits_T x^2 \sin \pi y\,dA\,.$$

Answer. The crucial question here is "What are the bounds of integration for the iterated integrals that correspond to this bound?" The first guess might be to have both

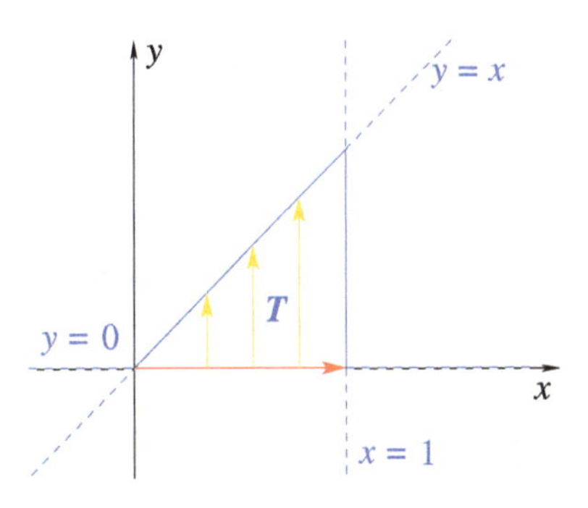

Figure 5.3: Triangular domain of integration T set off by solid blue segments, bounded by $y = x$, $x = 1$ and $y = 0$. The gold rays indicate how the y-integration depends on x: for each x-value corresponding to T, the ray begins at $y = 0$ and ends at $y = x$. The single red ray indicates the final x-integration from $x = 0$ to $x = 1$.

x- and y-integrals go from 0 to 1, but in fact we already know that these choices correspond to the unit square: $S := [0,1] \times [0,1]$. Notice that for the upper boundary, neither x or y is constant. Let us put the y-integral inside the x-integral; notice that the lower integration limit is $y = 0$ independent of x. The key observation about the upper limit of integration is that it depends on x: the upper limit is the line $y = x$. So the integration is

$$\iint_T x^2 \sin \pi y \, dA = \int_0^1 \left(\int_0^x x^2 \sin \pi y \, dy \right) dx = \int_0^1 x^2 \left(\int_0^x \sin \pi y \, dy \right) dx$$
$$= -\int_0^1 \frac{x^2}{\pi} \left(\cos \pi y \Big|_0^x \right. dx = \int_0^1 \frac{x^2}{\pi} (1 - \cos \pi x) \, dx = \frac{1}{3\pi} + \frac{2}{\pi^2} = \frac{\pi + 6}{3\pi^2} .$$

Notice that because the integrand was in fact the *product* of a function depending only on x and a function depending only on y, the x-function can be treated as a constant for the y-integration, and thus placed outside the y-integral. Also evaluating the final integral requires (a) double integration by parts, (b) a set of tables of integrals, or (c) a CAS (computer algebra system). The author recommends a computer.

Our next example shows that the y-integral should not always be evaluated first.

Example 5.7. Please evaluate the double integral

$$\iint_D 4xy^2 \, dA$$

where the domain of integration D is the bounded region between the x-axis, the curve $y = x^2$, and the line $y = 8 - 2x$ as shown in Figure 5.4,

Answer. Suppose that we try the typical first approach of carrying out the y-integration first. This approach is indicated by the gold rays in Figure 5.4. The small problem is that under this approach, the upper boundary of the domain changes as one carries out the integration: it changes from $y = x^2$ for $0 \leq x \leq 2$ to $y = 8 - 2x$ for $2 \leq x \leq 4$. This change means that to complete the y-integration first, one must break the iterated integral into two part, based on the two distinct upper boundary curves. Thus

$$\iint_D 4xy^2 \, dA = \int_0^2 \left(\int_0^{x^2} 4xy^2 \, dy \right) dx + \int_2^4 \left(\int_0^{8-2x} 4xy^2 \, dy \right) dx$$
$$= 4 \int_0^2 x \left(\frac{y^3}{3} \Big|_0^{x^2} \right. dx + 4 \int_2^4 x \left(\frac{y^3}{3} \Big|_0^{8-2x} \right. dx$$

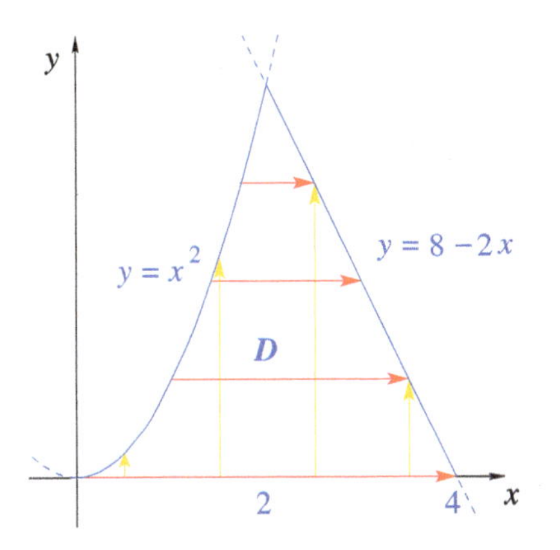

Figure 5.4: Domain of integration D bounded by the x-axis, the curve $y = x^2$, and the line $y = 8 - 2x$. Again the gold rays indicate how the y-integration depends on x, but now the red rays indicate how the x-integration depends on y.

$$
\begin{aligned}
&= \frac{4}{3}\int_0^2 x^7\,dx + \frac{4}{3}\int_2^4 512x - 384x^2 + 96x^3 - 8x^4\,dx \\
&= \frac{1}{6}x^8\Big|_0^2 + \frac{4}{3}\left(256x^2 - 128x^3 + 24x^4 - \frac{8}{5}x^5\Big|_2^4\right. \\
&= \frac{128}{3} + \frac{4}{3}\left(256(12) - 128(56) + 24(240) - \frac{8}{5}(992)\right) \\
&= \frac{128}{3} + \frac{4}{3}\left(1664 - \frac{7936}{5}\right) = \frac{2176}{15}.
\end{aligned}
$$

The above computation is simple but tedious, in part because there are two separate integrations. One might ask whether there might be an easier way? The answer is yes, though there is still some tedious work: carry out the x-integration first. This is indicated by the red rays in Figure 5.4.

$$
\begin{aligned}
\iint_D 4xy^2\,dA &= \int_0^4\left(\int_{\sqrt{y}}^{4-y/2} 4xy^2\,dx\right)dy = 4\int_0^4 y^2\left(\frac{x^2}{2}\Big|_{\sqrt{y}}^{4-y/2}\right.dy \\
&= 2\int_0^4 y^2\left((4-y/2)^2 - y\right)dy = \left(\frac{32}{3}y^3 - 2y^4 + \frac{y^5}{10}\Big|_0^4\right. \\
&= \frac{10240 - 9600 + 1536}{15} = \frac{2176}{15}.
\end{aligned}
$$

This second approach is surely less tedious; it is helped in part because of how the limits of integration interact with the integrand as the integration proceeds. If the integrand were different, this second approach could be more tedious.

In general, it is not always possible to predict which integration order is best when both are possible. One simply must try both orders to see which is easier. Of course, as was the case above, these to calculations can be used as a check on each other.

5.2.2 Polar integration

When the domain of integration D has polar rather than rectangular symmetry, the use of polar coordinates may be called for.

Example 5.8. Consider the semicircular disk D shown in Figure 5.5. Please compute

$$\iint_D x^2 y \, dA .$$

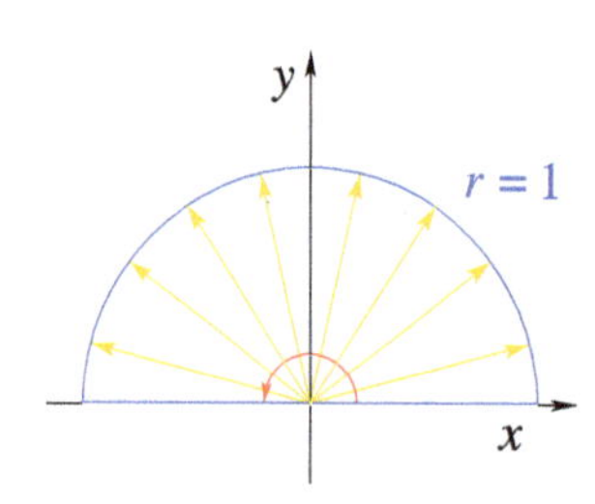

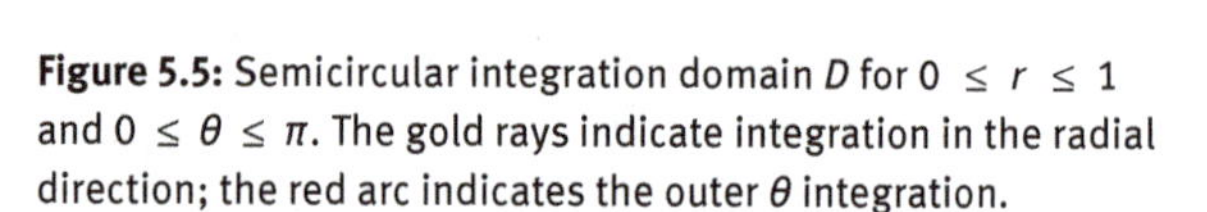

Figure 5.5: Semicircular integration domain D for $0 \le r \le 1$ and $0 \le \theta \le \pi$. The gold rays indicate integration in the radial direction; the red arc indicates the outer θ integration.

Answer. In this case, the limits of integration are constant in polar coordinates. For each θ, radial integration must go from $r = 0$ to $r = 1$. Then the angle θ must be integrated from $\theta = 0$ to $\theta = \pi$ to cover the entire semicircular disk. So in this regard, this integration is very much like integration over a rectangle in rectangular coordinates. Next, the integrand x^2y much written in polar coordinates: $x^2y = r^3 \cos^2\theta \sin\theta$.

It may seem that expressing the integration limits and the integrand in polar coordinates are the only two issues that need to be resolved to carry out polar integration, but there is one more: dA must be expressed in polar coordinates. In rectangular coordinates, dA in the double integral becomes $dx\,dy$ in the iterated integral, but in polar coordinates it is a bit more complicated: dA now becomes $r\,dr\,d\theta$. Why dA has this form is discussed after the example; the good news is that $r\,dr\,d\theta$ always replaces dA whenever a double integral is written in polar coordinates, no matter what the integration limits and integrand are.

Thus

$$\iint_D x^2 y \, dA = \int_0^{\pi} \left(\int_0^1 r^3 \cos^2\theta \sin\theta \, r dr \right) d\theta = \int_0^{\pi} \cos^2\theta \sin\theta \left(\frac{r^5}{5} \bigg|_0^1 \right) d\theta$$

$$= \frac{1}{5}\int_0^{\pi} \cos^2\theta\,(\sin\theta\,d\theta) = \frac{1}{5}\int_1^{-1} u^2\,(-du) = \frac{1}{5}\int_{-1}^{1} u^2\,du = \frac{2}{15}.$$

The above integration uses the substitution $u := \cos\theta$.

Notice that this integral can also be evaluated in rectangular coordinates, though the limits of integration are no longer constant:

$$\iint_D x^2 y\,dA = \int_{-1}^{1}\left(\int_0^{\sqrt{1-x^2}} x^2 y\,dy\right)dx = \int_{-1}^{1} x^2\left(\int_0^{\sqrt{1-x^2}} y\,dy\right)dx$$
$$= \int_{-1}^{1} x^2\left(\left.\frac{y^2}{2}\right|_0^{\sqrt{1-x^2}}\right)dx = \frac{1}{2}\int_{-1}^{1} x^2(1-x^2)\,dx$$
$$= \frac{1}{2}\int_{-1}^{1} x^2 - x^4\,dx = \frac{1}{2}\left(\frac{x^3}{3} - \left.\frac{x^5}{5}\right|_{-1}^{1}\right) = \left(\frac{1}{3} - \frac{1}{5}\right) = \frac{2}{15}.$$

Notice that the upper limit of integration is now the upper semicircle in rectangular coordinates: $y = \sqrt{1-x^2}$. Evaluating an integral in rectangular coordinates when the domain is easily described in polar coordinates may be much more difficult than it was in this example—indeed it might be impossible.

5.2.3 What does *dA* or *dy dx* become?

Why does the $dx\,dy$ of rectangular coordinates become $r\,dr\,d\theta$ in polar coordinates? An intuitive answer is shown in Figure 5.6. The key point here is that the length of the arc of a circle that subtends the angle $\Delta\theta$ is $r\,\Delta\theta$, thus increasing with r. But while Figure 5.6 is suggestive, it is not rigorous. What is needed is a theorem. The following result tells us how to change variables in $\mathbb{R}$. Its proof is not presented here, but can be found in Rudin [5, pp. 252–253]

Theorem 16. *If f is continuous on a bounded, connected domain $D \subset \mathbb{R}^2$ that can be described in either rectangular coordinates or a second coordinate pair ξ and η, then*

$$\iint_D f(x,y)\,dx\,dy = \iint_D f(x(\xi,\eta), y(\xi,\eta))\left|\frac{\partial(x,y)}{\partial(\xi,\eta)}\right| d\xi\,d\eta$$

where the Jacobian determinant is defined as

$$\frac{\partial(x,y)}{\partial(\xi,\eta)} := \begin{vmatrix} \frac{\partial x}{\partial\xi} & \frac{\partial x}{\partial\eta} \\ \frac{\partial y}{\partial\xi} & \frac{\partial y}{\partial\eta} \end{vmatrix}$$

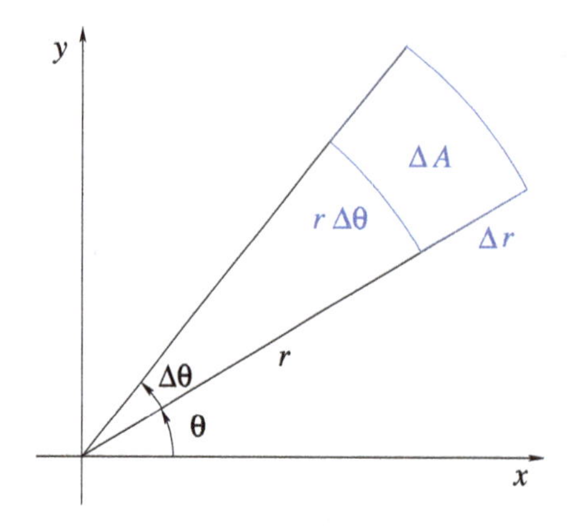

Figure 5.6: The area ΔA (in blue) of the portion of the $\mathbb{R}^2$-plane with radial length Δr subtending an angle $\Delta\theta$ at a distance r from the origin is $\Delta A \simeq (\Delta r)(r\,\Delta\theta) = r\,\Delta r\Delta\theta$. Notice that the plane can be tiled with patches of this form. This tiling suggests that in polar coordinates, dA should become $r\,dr\,d\theta$.

In the case of polar coordinates,

$$\frac{\partial(x,y)}{\partial(r,\theta)} = \begin{vmatrix} \frac{\partial x}{\partial r} & \frac{\partial x}{\partial \theta} \\ \frac{\partial y}{\partial r} & \frac{\partial y}{\partial \theta} \end{vmatrix} = \begin{vmatrix} \cos\theta & -r\sin\theta \\ \sin\theta & r\cos\theta \end{vmatrix} = r\cos^2\theta + r\sin^2\theta = r\,,$$

and this justifies the presence of a r when integrating in polar coordinates.

Remark. Careful inspection of the second integral in Theorem 16 might lead someone to think that there is a redundant determinant in this integral. In fact, this is not a determinant, it is an absolute value. In other words, it is indeed the absolute value of the determinant in the change of variable formula for multiple integrals.

The next example is a bit more difficult, but it still can be solved directly in polar coordinates.

Example 5.9. Let S be the portion of the circular disk $x^2 + (y-1)^2 \le 1$ lying in the first quadrant as shown in Figure 5.7. Please compute

$$\iint_S xy\,dA\,.$$

Answer. At first glance, it may seem that this domain would be difficult to describe in polar coordinates, at least without setting up a nonstandard polar system with its origin at the center of this disk. Standard polar coordinates, however, can be used to evaluate the integral in this example.

The key issue is finding the equation of the boundary circle for this disk in polar coordinates. Notice that in rectangular coordinates, this circle is $x^2 + (y-1)^2 = 1$ since the center is at $(0,1)$ and the radius is 1. Setting $x = r\cos\theta$ and $y = r\sin\theta$, one finds that $r^2 - 2r\sin\theta = 0$. Since normally $r > 0$ in polar coordinates, it must be the case

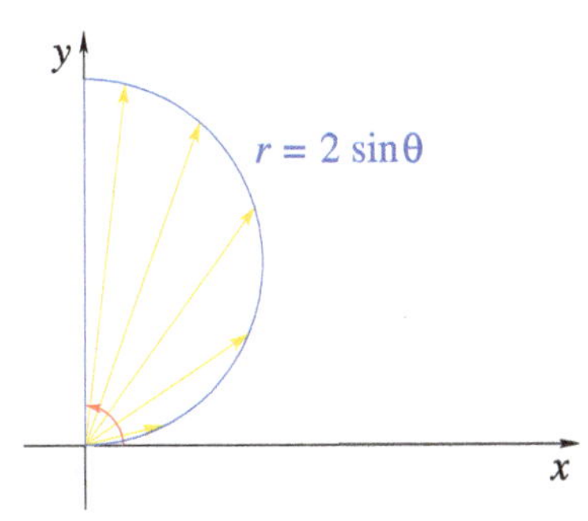

Figure 5.7: Semicircular disk S centered at $(0, 1)$ with radius 1 lying in the first quadrant. The gold rays indicate integration in the radial direction; the red arc indicates the outer θ integration.

that $r = 2\sin\theta$, and this is the equation of this circle in polar coordinates. Using this equation, one can see that integration over this circular disk can be described as in Figure 5.7, resulting in the following computation:

$$\iint_S xy\,dA = \int_0^{\pi/2}\left(\int_0^{2\sin\theta} r^2\cos\theta\sin\theta\,r dr\right)d\theta = \int_0^{\pi/2}\cos\theta\sin\theta\left(\frac{r^4}{4}\bigg|_0^{2\sin\theta}\right)d\theta$$
$$= 4\int_0^{\pi/2}\sin^5\theta\,(\cos\theta\,d\theta) = 4\int_0^1 u^5\,du = \frac{2}{3}.$$

The above integration uses the substitution $u := \sin\theta$, and again dA is represented in polar integration as $r\,dr\,d\theta$.

5.2.4 What does it all mean? What do double integrals represent?

Until now, our discussion of double integrals has concentrated on how to evaluate them using rectangular or polar coordinates. But equally important to evaluation is understanding how to interpret these integrals, that is, understanding what these integrals represent. As before, we can start by reviewing the single-variable case.

For a continuous, nonnegative function f defined on an interval $[a, b]$, in single-variable calculus, one learns that

$$\int_a^b f(x)\,dx = A$$

where A is the area under the curve $y = f(x)$ above the x-axis between a and b. But this area could also be found as a double integral, specifically,

$$\int_a^b\int_0^{f(x)} dy\,dx = \int_a^b\int_0^{f(x)} 1\,dy\,dx = \int_a^b f(x)\,dx = A$$

where the integrand in the first double integral is implicitly 1, implying that the y-integral can be evaluated immediately to reach the generic integral from single-variable calculus. This leads to the first major interpretation for the double integral: If $D \subset \mathbb{R}^2$ is a bounded with a piecewise smooth boundary ∂D, then

$$\iint_D dA = \iint_D 1\, dA = \text{Area}(D)\,.$$

A second generalization of the integral from single-variable calculus involves keeping f in the integrand, but changing the domain from $[a,b]$ to some bounded $D \subset \mathbb{R}^2$. Suppose again that the function $f \geq 0$ is continuous. This situation is depicted in Figure 5.8; from this figure, one can see that

$$\iint_D f(x,y)\, dA = V$$

where V is the volume under the surface $z = f(x,y)$ above the domain $D \subset \mathbb{R}^2$.

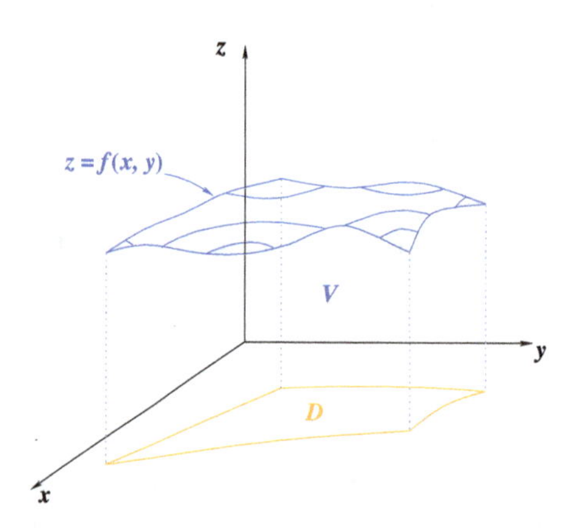

Figure 5.8: The volume V under the surface $z = f(x,y)$ above the domain $D \subset \mathbb{R}^2$.

Double integrals can also be used to define the average value of functions of two variables, as well as the center or centroid of a bounded, two-dimensional domain, D:

Definition. Let the area of D be

$$|D| \equiv \text{Area}(D) = \iint_D dA\,.$$

The *average value* of a function f on a domain D is defined as

$$f_{\text{av}}(D) := \frac{1}{|D|} \iint_D f(x,y)\, dA$$

and the *center* of D is $(\bar{x}, \bar{y})$ where

$$\bar{x} := \frac{1}{|D|} \iint_D x \, dA \quad \text{and} \quad \bar{y} := \frac{1}{|D|} \iint_D y \, dA$$

provided that all of the integrals above exist and are finite.

Example 5.10. Please find the center of the triangle T bounded by the x-axis, the line $y = x$ and the line $y = 6 - 2x$. Hint: Draw a diagram for the integration domain.

Answer. Because here the domain is a triangle, there are other ways to find its center, but using the definitions above, one finds that

$$|T| = \int_0^2 \left(\int_y^{3-y/2} dx \right) dy = 3$$

so

$$\bar{x} := \frac{1}{3} \int_0^2 \left(\int_y^{3-y/2} x \, dx \right) dy = 5/3 \quad \text{and} \quad \bar{y} := \frac{1}{3} \int_0^2 \left(\int_y^{3-y/2} y \, dx \right) dy = 2/3 .$$

One can compare these values with those found using the traditional definition for the center of a triangle to see that they are the same.

5.3 Triple integrals: integration over domains in $\mathbb{R}^3$

We now turn our attention to integration in three-dimensional space; as in the previous section, two of the most important issues in evaluating triple integrals are which coordinate system to use and in which order to integrate the variables.

5.3.1 Integration using rectangular coordinates

Consider the general triple integral

$$\iiint_\Omega f(x, y, z) \, dV$$

where the integrand f is continuous and the integration domain Ω is bounded, connected, and can be described as

$$\Omega = \{(x, y, z) \in \mathbb{R}^3 \mid a_1 \leq x \leq b_1, \ a_2(x) \leq y \leq b_2(x), \ a_3(x, y) \leq z \leq b_3(x, y)\}$$

where $z = a_3(x,y)$ and $z = b_3(x,y)$ are smooth surfaces defined on some domain $D = \{(x,y) \in \mathbb{R}^2 \mid a_1 \leq x \leq b_1, a_2(x) \leq y \leq b_2(x)\}$ and in turn, $y = a_3(x)$ and $y = b_3(x)$ are smooth curves defined on some interval $[a_1, b_1]$. Implicitly we assume that a_3 lies below b_3 (i. e., $a_3(x,y) \leq b_3(x,y)$ for all $(x,y) \in D$) and that a_2 lies below or to the left of b_2 (i. e., $a_2(x) \leq b_2(x)$ for all $x \in [a_1, b_1]$). This integration domain Ω is shown in Figure 5.9.

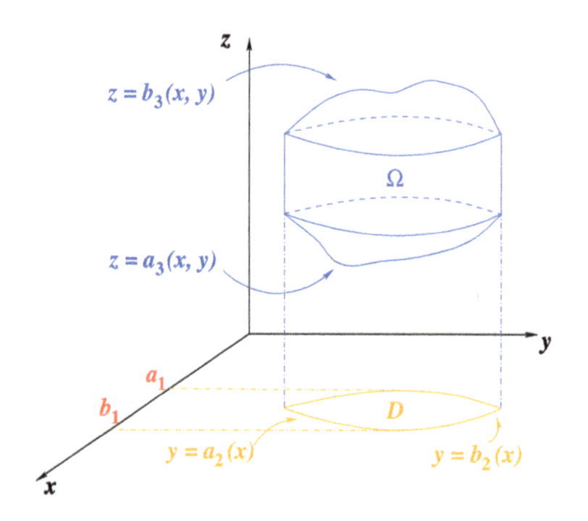

Figure 5.9: The integration domain $\Omega \subset \mathbb{R}^3$ where Ω lies above the surface $z = a_3(x,y)$ and below the surface $z = b_3(x,y)$. The projection of Ω onto the x,y-plane is D which is between the curves $y = a_2(x)$ and $y = b_2(x)$. Finally, the projection of D onto the x-axis is $[a_1, b_1]$.

As was the case for the double integral, the main tool for computing triple integrals is the Fubini theorem.

Theorem 17 (Fubini). *If f is continuous on a domain Ω described above, then*

$$\iiint_{\Omega} f(x,y,z)\,dV = \int_{a_1}^{b_1} \left(\int_{a_2(x)}^{b_2(x)} \left(\int_{a_3(x,y)}^{b_3(x,y)} f(x,y,z)\,dz \right) dy \right) dx\,.$$

Example 5.11. Please evaluate

$$\iiint_{C} 1 + x^2 + y^2 + z^2\,dV$$

where $C := [-1,1] \times [0,2] \times [1,3]$ is a cube (see Figure 5.10).

Answer. This requires, of course, just a direct application of the Fubini theorem:

$$\iiint_{C} 1 + x^2 + y^2 + z^2\,dV = \int_{-1}^{1} \left(\int_{0}^{2} \left(\int_{1}^{3} 1 + x^2 + y^2 + z^2\,dz \right) dy \right) dx$$

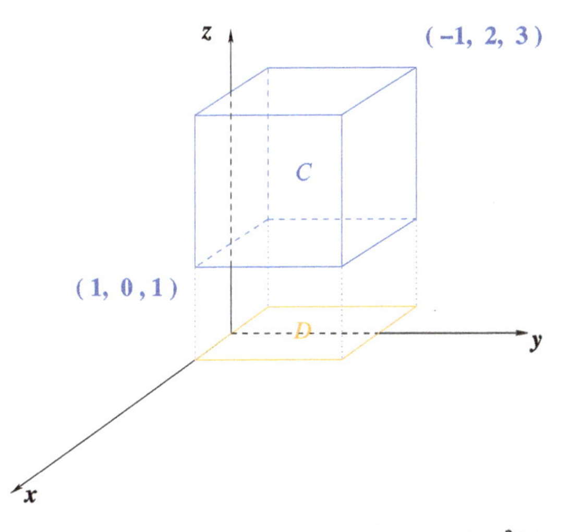

Figure 5.10: The cube $C := [-1,1] \times [0,2] \times [1,3]$ in $\mathbb{R}^3$ lying above the square $D := [-1,1] \times [0,2]$. Notice that this cube is the same whether the integration variables are x, y, and z, or x_1, x_2, and x_3.

$$= \int_{-1}^{1} \left(\int_{0}^{2} \left(z + x^2 z + y^2 z + \frac{z^3}{3} \Big|_1^{z=3} dy \right) dx = 2 \int_{-1}^{1} \left(\int_{0}^{2} \frac{16}{3} + x^2 + y^2 \, dy \right) dx\right.$$

$$= 2 \int_{-1}^{1} \left(\frac{16}{3} y + x^2 y + \frac{y^3}{3} \Big|_0^{y=2} dx = 4 \int_{-1}^{1} \frac{20}{3} + x^2 \, dx\right.$$

$$= 8 \int_{0}^{1} \frac{20}{3} + x^2 \, dx = 8 \left(\frac{20}{3} x + \frac{x^3}{3} \Big|_0^1 = \frac{168}{3}\right.$$

The simplest interpretation of a triple integral is as the volume of the integration domain Ω; in this case, the integrand is $f \equiv 1$ as in the next example.

Example 5.12. Please find the volume of a prism P in $\mathbb{R}^3$ which lies in the first octant and is bounded by the plane passing through the points $(a, 0, 0)$, $(0, b, 0)$, and $(0, 0, c)$ where $a, b, c > 0$ (see Figure 5.11).

Answer. First, recall that the general equation of a plane in three-dimensional space is $Ax + By + Cz = D$ where A, B, C, and D are constants, and that three noncolinear points uniquely determine a plane. Because of the specific form of these three points, the only plane that passes them is

$$\frac{x}{a} + \frac{y}{b} + \frac{z}{c} = 1 .$$

Solving for z, one finds that the upper surface is $z = c(1 - x/a - y/b)$; the lower surface for this prism is of course $z = 0$. Next, we must find the domain in the x, y-plane that

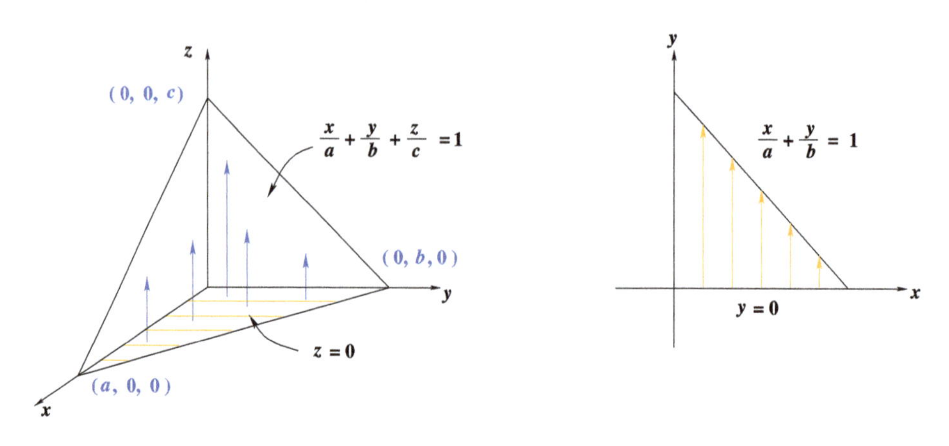

Figure 5.11: (Left) Prism P in the first octant bounded by the plane $(x/a) + (y/b) + (z/c) = 1$. The blue rays indicate the z-integration from $z = 0$ to the bounding plane (upper surface). (Right) The triangular domain in the x, y-plane that the prism lies above. Here, the gold rays indicate y-integration from $y = 0$ to $y = b(1 - x/a)$.

this prism sets above; in this case, this domain is the triangular base of the prism (see Figure 5.11). The diagonal line that bounds this triangular base in the first quadrant is defined by the intersection of the upper and lower surfaces: $y = b(1 - x/a)$. Setting up the double integral over this triangular base is then the same as it was in the previous section; assuming that integration in the y-direction is carried out first, one must integrate from the lower curve $y = 0$ to the upper curve $y = b(1 - x/a)$, then finally integrate across all x-values associated with the domain, from $x = 0$ to $x = a$. So the integration is

$$\iiint_P 1\,dV = \int_0^a \left(\int_0^{b(1-x/a)} \left(\int_0^{c(1-x/a-y/b)} dz \right) dy \right) dx = c \int_0^a \left(\int_0^{b(1-x/a)} 1 - \frac{x}{a} - \frac{y}{b}\,dy \right) dx$$
$$= c \int_0^a \left(y - \frac{xy}{a} - \frac{y^2}{2b} \Big|_0^{y=b(1-x/a)} dx = bc \int_0^a 1 - \frac{x}{a} - \frac{x}{a}\left(1 - \frac{x}{a}\right) - \frac{1}{2}\left(1 - \frac{x}{a}\right)^2 dx$$
$$= \frac{abc}{6}.$$

Domains can be more complicated than this first prism. The next example deals with a prism whose base is not on the x, y-plane. The next subsection deals with domains that are complicated to describe in rectangular coordinates, but much easier to describe in either cylindrical or spherical coordinates.

Example 5.13. Please find the volume V of the prism P in the first octant bounded above by the plane $z = 5 - 2x - y$ and below by the plane $z = 3x + 4y$ (see Figure 5.12).

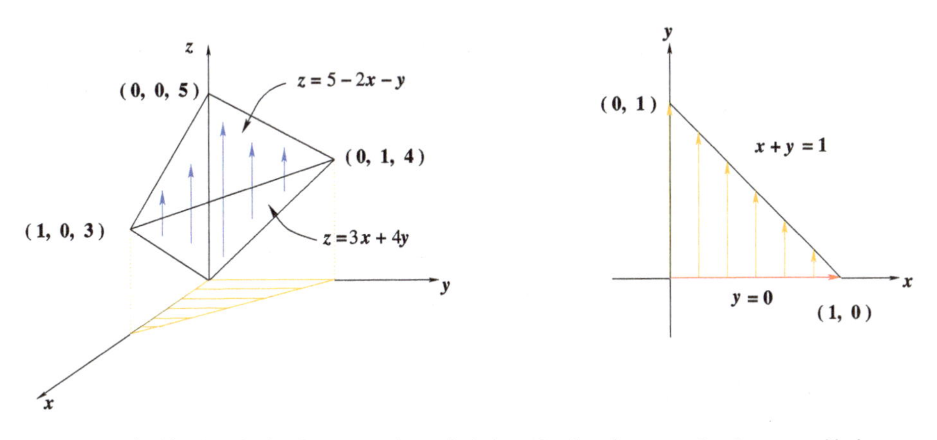

Figure 5.12: (Left) Prism in the first octant bounded above by the plane $z = 5 - 2x - y$ and below by the plane $z = 3x + 4y$. The blue rays indicate the z-integration from the lower bounding plane to the upper bounding plane. (Right) The triangular domain in the x, y-plane that the prism lies above. Here, the gold rays indicate y-integration from $y = 0$ to $y = 1 - x$.

Answer. Perhaps the most important thing in setting up and evaluating integrals over more complicated domains is an accurate drawing of the domain(s) involved. Sometimes such drawings are given, but if not, then they must be sketched. Here, the domains are shown in Figure 5.12. In this case, from both the diagram on the left in Figure 5.12 and the problem statement, it seems clear that the inner integration can go vertically from the lower surface $z = 3x + 4y$ to the upper surface $z = 5 - 2x - y$. The more complicated issue, perhaps, is what should be the integration limits for the other two dimensions—the x and y limits. The question here is again "Over which portion of the x, y-plane does this prism lie above?" Notice that in this case, the outer edge of this prism is defined as the intersection of the upper and lower planes. This intersection is found, of course, by setting equal the z values for both planes. The result in terms of x and y is that $x + y = 1$. This is shown on the right side of Figure 5.12: it is the triangular region bounded by the x and y axes and the line $y = 1 - x$. Here, the integration is

$$V = \iiint_P 1\,dV = \int_0^1 \left(\int_0^{1-x} \left(\int_{3x+4y}^{5-2x-y} dz \right) dy \right) dx = \int_0^1 \left(\int_0^{1-x} 5 - 5x - 5y\,dy \right) dx$$

$$= 5 \int_0^1 \left(y - xy - \frac{y^2}{2} \Big|_0^{y=1-x} \right) dx = \frac{5}{2} \int_0^1 (1-x)^2\,dx = \frac{5}{6}.$$

What should one do if a drawing of the integration domain is not given? Often the answer is that it must be drawn, perhaps by hand, perhaps using a computer. But sometimes it is possible to find the integration limits without an actual drawing of Ω.

Example 5.14. Please find the volume of the object Ω bounded by the surfaces $z = x^2 + y^2$, $y = x^2$, $y = 9$, $z = \sin \pi x$, and $x = 1$.

Answer. At first glance, this may seem like a mess of equations—and a mess of surfaces. But notice that three of these equations involve only x and y, not z, so they bound a domain D in the x,y-plane. Considering these three equations, and perhaps making a quick sketch of them in the x,y-plane, one finds that integration limits in y are from $y = x^2$ to $y = 9$, and in x from $x = 1$ to $x = 3$. The two remaining equations both involve the third variable z; notice that based on the domain D in the x,y-plane, the surface $z = x^2 + y^2 > 2$ on D, while $z = \sin \pi x \leq 1$. This tells us which is the upper surface and which is the lower. So the integration is

$$V = \iiint_{\Omega} 1\, dV = \int_1^3 \left(\int_{x^2}^{9} \left(\int_{\sin \pi x}^{x^2+y^2} dz \right) dy \right) dx$$
$$= \int_1^3 \left(\frac{x^4}{3} + 4x^2 + 27 - \sin \pi x \right) (9 - x^2)\, dx = \frac{45686}{105} + \frac{8}{\pi} \simeq 437.65\,.$$

Notice that several integrals involving $\sin \pi x$ must be zero because of its symmetry on $[1,3]$.

5.3.2 Integration using cylindrical and spherical coordinates

When the domain of integration and the integrand have rectangular symmetry (or no symmetry at all), using rectangular coordinates in setting up and evaluating integrals makes perfect sense. But what happens when a problem involves cylindrical or spherical symmetry? Can one use those coordinate systems to simplify integration? The answer of course is yes, but to take advantage of either of these systems, one must be able to express dV in one or both of these coordinate systems.

5.3.2.1 Cylindrical coordinates

Expressing dV in cylindrical coordinates can be done using Theorem 16, but here we take advantage of cylindrical coordinates being a combination of polar coordinates for the x,y-plane and rectangular coordinates in the vertical z direction. Thus provided that the Fubini theorem can be applied,

$$\iiint_{\Omega} f(x,y,z)\, dV = \iiint_{\Omega} f(x,y,z)\, dz\, dy\, dx = \iint_D \left(\int_{c(x,y)}^{d(x,y)} f(x,y,z)\, dz \right) dA$$
$$= \iint_D \left(\int_{c(x,y)}^{d(x,y)} f(x,y,z)\, dz \right) r\, dr\, d\theta = \iiint_{\Omega} f(r\cos\theta, r\sin\theta, z)\, dz\, r\, dr\, d\theta$$

where the integration domain Ω lies between the surfaces $z = c(x,y)$ (below) and $z = d(x,y)$ (above) and projects onto D in the x,y-plane. In the above relation, each equality is based on a version of the Fubini theorem. A heuristic drawing which suggests why dV in a triple integral corresponds to $dz\,r\,dr\,d\theta$ for an iterated integral in cylindrical coordinates is shown in Figure 5.13.

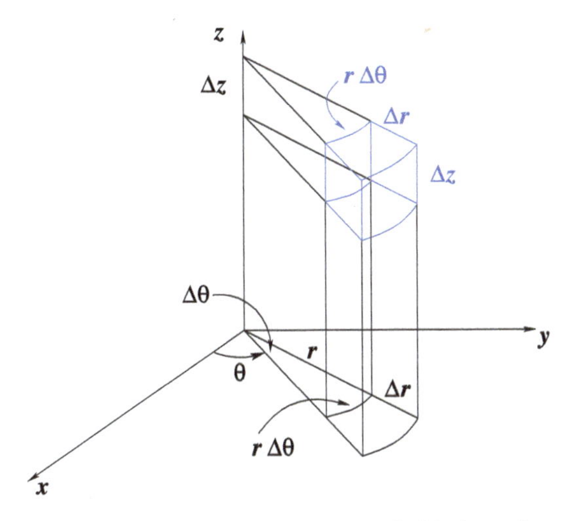

Figure 5.13: Schematic showing ΔV in cylindrical coordinates. Here, the polar area ΔA shown in Figure 5.6 is projected in the vertical z direction to a thickness Δz to form the cylindrical ΔV shown in blue. $\Delta V = (\Delta z)(\Delta r)(r\Delta\theta) \Rightarrow dV \to dz\,r dr\,d\theta$.

Example 5.15. Please compute

$$\iiint_\Omega 1 + x\,dV$$

where Ω is the cylinder bounded by the circle $x^2 + y^2 = 16$ and the planes $z = 3$ and $z = 7$.

Answer. It seems clear that nothing has more cylindrical symmetry than a cylinder, so the integration domain begs for the use of cylindrical coordinates. The integrand $1 + x$ does not have cylindrical symmetry, but this would seem to be a small price to pay for the benefit of having constant integration limits with z running from 3 to 7, r from 0 to 4, and θ from 0 to 2π. Carrying out the integration, one finds that

$$\iiint_\Omega 1 + x\,dV = \int_0^{2\pi}\left(\int_0^4\left(\int_3^7 1 + r\cos\theta\,dz\right) r\,dr\right) d\theta$$

$$= 4\int_0^{2\pi}\left(\int_0^4 r + r^2\cos\theta\,dr\right) d\theta = 4\int_0^{2\pi}\left(\frac{r^2}{2} + \frac{r^3}{3}\cos\theta\Big|_{r=0}^{r=4}\right. d\theta$$

$$= 64 \int_0^{2\pi} \left(\frac{1}{2} + \frac{4}{3} \cos\theta \right) d\theta = 64(\pi + 0) = 64\pi \,.$$

Notice that the second part of the integral being zero could have been anticipated since x has odd symmetry relative to this cylinder (it equally takes on positive and negative values on halves of the cylinder).

Example 5.16. Please find the volume of the parabolic dome P bounded above by the paraboloid $z = 8 - x^2 - y^2$ and below by the x, y-plane (see Figure 5.14).

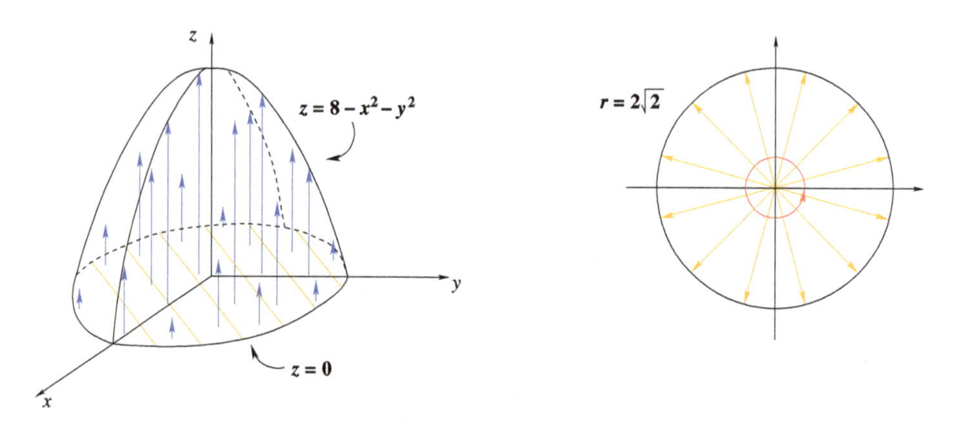

Figure 5.14: The parabolic dome bounded above by the paraboloid $z = 8 - x^2 - y^2$ and below by the x, y-plane.

Answer. Even though this dome is not a cylinder, it does have a large degree of cylindrical symmetry, suggesting that cylindrical coordinates are appropriate. As indicated in Figure 5.14, the inner-most z-integration goes from $z = 0$ up to the paraboloid $z = 8 - x^2 - y^2 = 8 - r^2$. Integration in the x, y-plane is then over the base of this parabolic dome; the outer edge of this base is the circle defined by the paraboloid $z = 8 - r^2$ intersecting the x, y-plane where $z = 0$:

$$z = 0 = 8 - r^2 \quad \Rightarrow \quad r = 2\sqrt{2}$$

Thus the radial integration goes from $r = 0$ to $r = 2\sqrt{2}$, and the angular integration goes once around the plane from $\theta = 0$ to $\theta = 2\pi$. The integration is then

$$\iiint_P dV = \int_0^{2\pi} \left(\int_0^{2\sqrt{2}} \left(\int_0^{8-r^2} dz \right) r\, dr \right) d\theta$$

$$= 2\pi \int_0^{2\sqrt{2}} (8 - r^2)\, r\, dr = 2\pi \left(4r^2 - \frac{r^4}{4} \right|_0^{2\sqrt{2}} = 2\pi(32 - 16) = 32\pi \,.$$

Notice that since the integration limits in r and z and the integrand itself do not depend on θ, the outer θ integral may be evaluated first. Indeed this is frequently the case for cylindrical integration—one often can immediately find a factor of 2π in front of the integration.

5.3.2.2 Spherical coordinates

To use spherical coordinates in integration, dV must be expressed in spherical coordinates, and this can be done using the version of Theorem 16 for integration domains in $\mathbb{R}^3$:

$$\iiint_{\Omega} f(x,y,z)\,dx\,dy\,dz = \iiint_{\Omega} f(\rho\sin\phi\cos\theta, \rho\sin\phi\sin\theta, \rho\cos\phi)\left|\frac{\partial(x,y,z)}{\partial(\rho,\phi,\theta)}\right| d\rho\,d\phi\,d\theta$$

where the Jacobian determinant in this case (see Exercise 5.14) is

$$\frac{\partial(x,y,z)}{\partial(\rho,\phi,\theta)} = \begin{vmatrix} \frac{\partial x}{\partial\rho} & \frac{\partial x}{\partial\phi} & \frac{\partial x}{\partial\theta} \\ \frac{\partial y}{\partial\rho} & \frac{\partial y}{\partial\phi} & \frac{\partial y}{\partial\theta} \\ \frac{\partial z}{\partial\rho} & \frac{\partial z}{\partial\phi} & \frac{\partial z}{\partial\theta} \end{vmatrix} = \rho^2\sin\phi\,.$$

Thus the transformation of an integral from rectangular to spherical coordinates is

$$\iiint_{\Omega} f(x,y,z)\,dx\,dy\,dz = \iiint_{\Omega} f(\rho\sin\phi\cos\theta, \rho\sin\phi\sin\theta, \rho\cos\phi)\rho^2\sin\phi\,d\rho\,d\phi\,d\theta\,.$$

Intuitively, one may arrive at this change of variable identity as is shown in Figure 5.15.

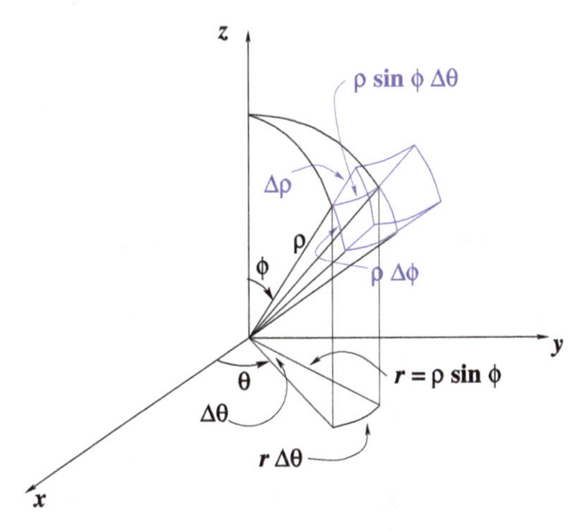

Figure 5.15: Schematic showing ΔV in spherical coordinates. Note that unlike in previous cases, none of the edges of ΔV (shown in blue) are parallel here. $\Delta V = (\rho\Delta\phi)(\Delta\rho)(\rho\sin\phi\Delta\theta) \Rightarrow dV \to \rho^2\sin\phi\,d\rho\,d\phi\,d\theta$.

The above formula makes spherical integration seem complicated; in fact, when the integration domain and integrand have spherical symmetry, the use of spherical coordinates greatly simplifies our work as the next example makes clear.

Example 5.17. What is the volume of a sphere of radius R?

Answer. For simplicity, let us call the sphere S, its volume, V, and assume that it is centered at the origin $(0, 0, 0)$. Putting the center at the origin is a matter of convenience; the volume would be the same no matter where it is centered. Then

$$\begin{aligned} V &= \iiint_S dV = \iiint_S \rho^2 \sin\phi \, d\rho \, d\phi \, d\theta \\ &= \int_0^{2\pi} \left(\int_0^{\pi} \left(\int_0^{R} \rho^2 \sin\phi \, d\rho \right) d\phi \right) d\theta = 2\pi \frac{R^3}{3} \int_0^{\pi} \sin\phi \, d\phi = \frac{4}{3}\pi R^3 . \end{aligned}$$

This is of course the famous formula for the volume of a sphere. It can be found in a number of other ways (see Exercise 5.15), including integrating in rectangular coordinates, but this would be much more complicated. Whenever possible, it is best to take advantage of symmetry.

The next example introduces another interpretation of a triple integral: mass. To obtain the mass M of a object Ω, one must integrate the mass density function δ over the entire volume (or object) Ω. In physical terms, mass density has units of mass per unit volume, so this integration in effect multiplies mass per unit volume by volume to obtain mass:

$$M = \iiint_\Omega \delta(x, y, z) \, dV .$$

The next example also makes clear that seeing symmetry may be difficult. Still there is often a payoff in terms of computational simplicity to taking advantage of the symmetry whenever possible.

Example 5.18 (Ice cream cone). Please find the mass M of an ice cream cone, C, bounded below by the cone $z^2 = 3(x^2+y^2)$, bounded above by the sphere $x^2+y^2+z^2 = 1$, if density is proportional to z, and the entire ice cream cone is above $z = 0$ (see Figure 5.16).

Answer 1. Since the density is proportional to z, we have that $\delta(x, y, z) = \delta_0 z$ for some constant δ_0. Because the density is proportional to z, and because of the geometry of the cone itself, cylindrical coordinates are the obvious choice for this integral. So the innermost z integral goes from the lower surface to the upper surface, and the outer double integral in polar coordinate must cover the disk in the x, y-plane that the ice cream cone stands above. The radius of this disk is determined by the intersection of

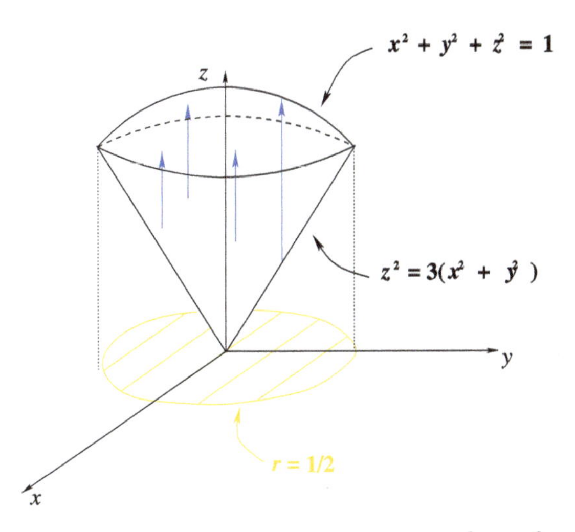

Figure 5.16: The ice cream cone bounded below $z^2 = 3(x^2 + y^2)$ and above by $x^2 + y^2 + z^2 = 1$. The blue rays indicate integration from the lower surface to the upper surface, while the gold disk in the x, y-plane has radius $r = 1/2$ and is the domain for polar integration.

the sphere above and the cone below: $z^2/3 = r^2 = 1 - z^2$ implying that $z = \sqrt{3}/2$ and $r = 1/2$. Finally, the integration is

$$M = \iiint_C \delta_o z\,dV = \int_0^{2\pi}\left(\int_0^{1/2}\left(\int_{\sqrt{3}r}^{\sqrt{1-r^2}} \delta_o z\,dz\right) r\,dr\right) d\theta = 2\pi\delta_o \int_0^{1/2}\left(\frac{z^2}{2}\bigg|_{\sqrt{3}r}^{\sqrt{1-r^2}}\right) r\,dr$$

$$= \pi\delta_o \int_0^{1/2} \left(r - 4r^3\right) dr = \pi\delta_o \left(\frac{1}{2}r^2 - r^4\right)\bigg|_0^{1/2} = \pi\delta_o/16\,.$$

The θ-integral may be evaluated immediately because neither of the other two integrals nor the integrand depend on θ.

Answer 2. Although it is less obvious, this cone and integrand also have spherical symmetry. Indeed all of the integration limits for this cone are constant in spherical coordinates. The ρ-integral must go from the origin $\rho = 0$ to the sphere $\rho = 1$; the φ-integral must go from the positive z-axis to the cone $\varphi = \pi/6$; finally, the θ-integral must go from 0 to 2π (see Figure 5.17). So in spherical coordinates the integration is

$$M = \iiint_C \delta_o z\,dV = \int_0^{2\pi}\int_0^{\pi/6}\int_0^{1} \delta_o \rho\cos\varphi\,\rho^2\sin\varphi\,d\rho\,d\varphi\,d\theta$$

$$= 2\pi\delta_o\left(\int_0^1 \rho^3\,d\rho\right)\left(\int_0^{\pi/6}\cos\varphi\sin\varphi\,d\varphi\right) = 2\pi\delta_o\left(\frac{1}{4}\right)\left(\frac{1}{8}\right) = \pi\delta_o/16$$

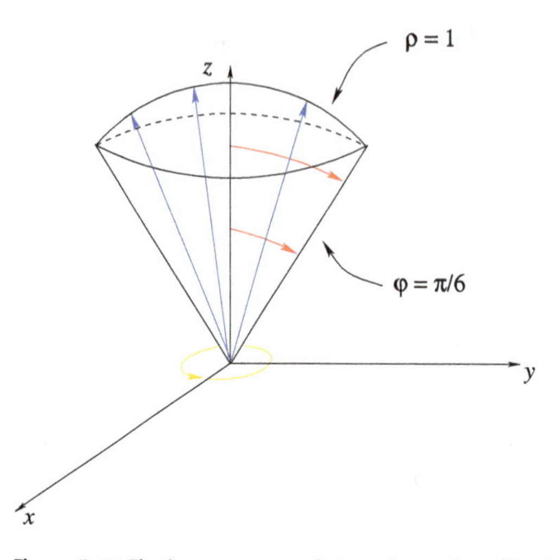

Figure 5.17: The ice cream cone integration using spherical coordinates. ρ-integration is indicated by blue rays from the origin to the sphere $\rho = 1$; φ-integration is indicated by red arc rays from the z-axis to the cone $\varphi = \pi/6$; and θ-integration is indicated by a gold loop starting and ending at the positive x-axis.

In this case, the triple iterated integral can be factored and written as the product of three separate integrals because all the limits of integration are constant, and the integrand factors into the product of three expressions, the first depending only on ρ, the second, only on φ, and the third, only on θ.

For three-dimensional objects, there is a distinction between the center, centroid, or geometric center[5] on the one hand, and the center of mass or center of gravity on the other.

Definition. For an object (domain) $\Omega \subset \mathbb{R}^3$, suppose that its density at any point $(x, y, z) \in \Omega$ is given by $\delta(x, y, z) \geq 0$. Recall that the volume of Ω be

$$|\Omega| \equiv \text{Volume}(\Omega) = \iiint_{\Omega} dV\,,$$

and the mass of the object is

$$M \equiv \text{Mass}(\Omega) = \iiint_{\Omega} \delta(x, y, z)\, dV\,.$$

The *average value* of a function f on a domain Ω is defined as

$$f_{\text{av}}(\Omega) := \frac{1}{|\Omega|} \iiint_{\Omega} f(x, y, z)\, dV\,,$$

5 Some authors use the terms centroid and center of mass interchangeably, but this leads to confusion.

the *center, centroid,* or *geometric center* of Ω is $(\bar{x}, \bar{y}, \bar{z})$ where

$$\bar{x} := \frac{1}{|\Omega|} \iiint\limits_{\Omega} x \, dV\,, \quad \bar{y} := \frac{1}{|\Omega|} \iiint\limits_{\Omega} y \, dV \quad \text{and} \quad \bar{z} := \frac{1}{|\Omega|} \iiint\limits_{\Omega} z \, dV\,,$$

and the *center of mass* of Ω is (x_m, y_m, z_m) where

$$x_m := \frac{1}{M} \iiint\limits_{\Omega} x\delta(x,y,z) \, dV\,, \quad y_m := \frac{1}{M} \iiint\limits_{\Omega} y\delta(x,y,z) \, dV$$

and

$$z_m := \frac{1}{M} \iiint\limits_{\Omega} z\delta(x,y,z) \, dV$$

provided that the appropriate integrals above exist and are finite and there is no division by zero.

Remarks.

1. Notice that if density is constant, say $\delta(x,y,z) \equiv \delta_o > 0 \; \forall \, x,y,z \in \Omega$, then the center of mass and the centroid are the same point.
2. These definitions can also be made for two-dimensional objects, and these same sorts of definitions can be made for densities other than mass density, for example, population density.

Example 5.19. Please find the center of mass of the ice cream cone, C, in Example 5.18: again C is bounded below by the cone $z^2 = 3(x^2 + y^2)$, bounded above by the sphere $x^2 + y^2 + z^2 = 1$, density is proportional to z, and the entire ice cream cone is above $z = 0$.

Answer. Notice that because of the symmetry of the cone and its density, $(x_m, y_m) = (0,0)$ and only z_m needs to be computed. From Example 5.18, the mass of this ice cream cone is $\pi\delta_o/16$, so

$$z_m = \frac{16}{\pi\delta_o} \iiint\limits_{C} z^2 \delta_o \, dV\,.$$

As before, one has the choice of using either cylindrical or spherical coordinates. In spherical coordinates, this integral becomes

$$z_m = \frac{16}{\pi}(2\pi)\left(\int_0^1 \rho^4 \, d\rho\right)\left(\int_0^{\pi/6} \cos^2\varphi \sin\varphi d\varphi\right) = 32\left(\frac{1}{5}\right)\left(\frac{8 - 3\sqrt{3}}{24}\right) = \frac{32 - 12\sqrt{3}}{15}\,.$$

The integrals in the expressions for the centroid and the center of mass have names in their own right.

Definition. The *first moment about the xy-plane* is

$$M_{xy} := \iiint\limits_{\Omega} z\, dV\,,$$

the *first moment about the xz-plane* is

$$M_{xz} := \iiint\limits_{\Omega} y\, dV\,,$$

and the *first moment about the yz-plane* is

$$M_{yz} := \iiint\limits_{\Omega} x\, dV\,.$$

Notice that z is the signed distance between the point (x, y, z) and the xy-plane, and this explains its presence in the definition of M_{xy}. The analogous statement is true for M_{xz} and M_{yz}. So $\bar{x} = M_{yz}/|\Omega|$, $\bar{y} = M_{xz}/|\Omega|$ and $\bar{z} = M_{xy}/|\Omega|$.

Remarks.
1. These definitions for moments are given based on the centroid rather than the center of mass; some authors would include the density in each definition, and thus based their definitions on the center of mass. Again, one must simply know the choices a given author has made.
2. The name "first moment" suggests that there is a second moment, and indeed this is the case. The second moment is the moment of inertia and is normally defined relative to an axis, rather than a plane. The moment of inertia for an object Ω is important in discussing angular momentum, and about the z-axis is defined as

   $$I_z := \iiint\limits_{\Omega} (x^2 + y^2)\, \delta(x, y, z)\, dV\,.$$

   Notice that for the moment of inertia one expects to see the density included. Also notice that $x^2 + y^2$ is the square of the distance from the point (x, y, z) to the z-axis.

Exercises 5

5.1. Following Example 5.1, please use a uniform partition with the left endpoint of each subinterval as the sampling point to compute

$$\int\limits_{0}^{4} x^3 dx$$

as the limit of Riemann sums. Use standard integration to confirm your answer.

Hint: You should confirm that here $\Delta x = 4/n$ and $\xi_i = 4i/n$. You may need to look up the formula for the sum of the first n integers cubed.

5.2. For the rectangle $R = [3,5] \times [0,3]$, please evaluate

$$\iint_R f(x,y)\,dA$$

for each of the following integrands, $f(x,y)$:
(a) $f(x,y) = 2x + 5x^2y$
(b) $f(x,y) = y\cos x$
(c) $f(x,y) = xe^{xy}$

Answer. (a) 783, (c) $(e^{15} - e^9)/3 - 2$

5.3. Please evaluate the following integral and sketch the integration domain:

$$\int_0^2 \left(\int_{3x-1}^{5x^2+2} x\,dy \right) dx$$

5.4. Suppose that f is continuous on an integration domain D. Please find the limits of integration for an iterated integral that is equivalent by the Fubini theorem to

$$\iint_D f(x,y)\,dA$$

for each of the following integration domains. You may choose the most convenient integration order and variables. **Hint:** A sketch may help.
(a) D is the triangle bounded by the y-axis, the line $y = 3$, and the line $y = 3x$.
(b) D is the disk sector in the first quadrant bounded by the x-axis, the line $y = x$, and the circle $x^2 + y^2 = 5$.
(c) D is the domain bounded by the line $y = 2x$, and the parabola $x = y^2$.
(d) D is the triangle bounded by the x-axis, the line $y = 2x$, and the line $y = 4x - 2$.
(e) D is the semicircular disk in the first quadrant bounded by the y-axis, and the circle $x^2 + y^2 = 4y$.

5.5. For the following integral, sketch the integration domain, and reverse the order of integration (i. e., put the x-integration on the inside and the y-integration on the outside):

$$\int_0^1 \left(\int_x^{\sqrt{x}} f(x,y)\,dy \right) dx$$

5.6. What is the area of the portion of the first quadrant bounded by the lemniscate $r = \sin 2\theta$. **Hint:** A sketch may help.

Answer. $\pi/8$

5.7. Please compute the center of each domain D:

(a) D is the triangle bounded by the y-axis, the line $y = 3$, and the line $y = 3x$.

(b) D is the semicircular disk bounded above by the unit circle and below by the x-axis. (Notice that by symmetry, $\bar{x} = 0$, so only $\bar{y}$ must be computed.)

(c) D is bounded by parabola $y = x^2 - 2x$ and the line $y = x$.

5.8. For the function $f(x, y) = \sqrt{x^2 + y^2}$,

(a) What is its average value on the unit disk $x^2 + y^2 \leq 1$?

(b) What is its average value on the annulus (washer) $1/2 \leq x^2 + y^2 \leq 1$?

Answer. (a) 2/3, (b) 7/9

5.9. Please evaluate

$$\iiint_C 1 + xyz \, dV$$

where $C := [1, 3] \times [2, 4] \times [3, 5]$ is a cube.

5.10. What is the volume of the spherical cap above the plane $z = 1$ and below the sphere $x^2 + y^2 + z^2 = 4$.

5.11. Please compute the volume of the solid paraboloid bounded below by $z = x^2 + y^2$ and above by the plane $z = 5$.

Answer. $25\pi/2$

5.12. Compute

$$\iiint_\Omega x + y^2 \, dV$$

where Ω is the solid bounded by surfaces $y = x^2$, $z = y$, $x = 0$, $z = 0$, and $y = 4$.

5.13. What is the volume of the prism in the first octant below the plane $z = 6 - x - 2y$ and above the plane $z = 2x + y$.

Answer. 4

5.14. Please compute the Jacobian determinant for the integration change of variables from rectangular to spherical coordinates discussed in Section 5.3.2.2 above. **Hint:** Expand by the first row, and take out front the common $\rho^2 \sin\phi$ factor.

5.15. Using cylindrical coordinates, please confirm that the volume of a sphere of radius R is $\frac{4}{3}\pi R^3$. Example 5.17 shows this computation in spherical coordinates.

5.16. Find the volume of the discus bounded above by $x^2 + y^2 + z^2 = 1$ and below by $x^2 + y^2 + (z - 3)^2 = 8$.

5.17. Please find the mass M of the bounded paraboloid where $x^2 + y^2 \le z \le 1$ if the density is proportional to $1 - z$.

Answer. $\pi\delta_o/3$ where δ_o is the proportionality constant for density.

5.18.

(a) Find the center of mass of the ice cream cone in Example 5.18 using cylindrical coordinates.

(b) Find the centroid of the ice cream cone in Example 5.18.

5.19. Please explain why, if density is a positive constant, then the centroid and the center of mass are the same point.

5.20. Compute the moment of inertia about the z-axis, I_z, for a solid sphere centered at the origin of radius R assuming that the density is constant.

Answer. $8\pi R^5/15$

6 Vector fields and vector calculus

This chapter deals with vector fields which can be thought of as multivariable vector functions. Here the domain is a subset of $\mathbb{R}^n$, and the range consists of vectors in $\mathbb{R}^n$. Typically the dimension for both the domain and the range are the same. The name "vector field" comes from the fact that when $n = 2$, a typical drawing of a vector field is as a field of vectors in the x, y-plane.

6.1 Line integrals: integration along curves in $\mathbb{R}^2$ or $\mathbb{R}^3$

The first thing that everyone should know about line integrals is that they only occasionally involve integrating over lines. Rather these integrals are generally over curves in $\mathbb{R}^2$ or $\mathbb{R}^3$, but somehow the term "curve integral" has never taken root. They are sometime called *contour integrals*, but this term is more common for integration in the complex plane rather than $\mathbb{R}^n$.

Definition. Let C be a smooth curve in $\mathbb{R}^n$ traced out by a differentiable vector function $\boldsymbol{x} : [a, b] \to \mathbb{R}^n$ with $\boldsymbol{x}(a)$ being the starting point α and $\boldsymbol{x}(b)$ being the ending point ω. Let $\boldsymbol{F} : D \subset \mathbb{R}^n \to \mathbb{R}^n$ be a vector field whose domain D is an open set containing C. Then the *line integral*

$$\int_C \boldsymbol{F}(\boldsymbol{x}) \cdot d\boldsymbol{x} := \int_a^b \boldsymbol{F}(\boldsymbol{x}(t)) \cdot \frac{d\boldsymbol{x}}{dt}\, dt \equiv \int_a^b \boldsymbol{F}(\boldsymbol{x}(t)) \cdot \dot{\boldsymbol{x}}(t)\, dt$$

provided that the integrals on the right of := exist and are finite.

Remarks.
1. In what follows, $n = 2$ or $n = 3$, though the definition above is the same for larger integer values of $n \in \mathbb{Z}^+$.
2. The line integral above represents or measures the tendency of the vector field $\boldsymbol{F}$ to flow *along* the curve C from the beginning point α to the ending point ω. Later we will discuss a separate type of line integral that represents or measures the tendency of the vector field $\boldsymbol{F}$ to flow *across* the curve C. See Figure 6.1.
3. In this definition, "smooth" can be replaced by "piecewise smooth" and then the definition can be applied separately to each piece. See Proposition 7 and Exercise 6.2 below.

There are a couple of basic results about line integrals that will be used frequently in this chapter. From the definition above, one might wonder if the value of a line integral can depend on how the curve C is parameterized. As one would hope, this is not the case, as the following proposition states.

https://doi.org/10.1515/9783110660609-006

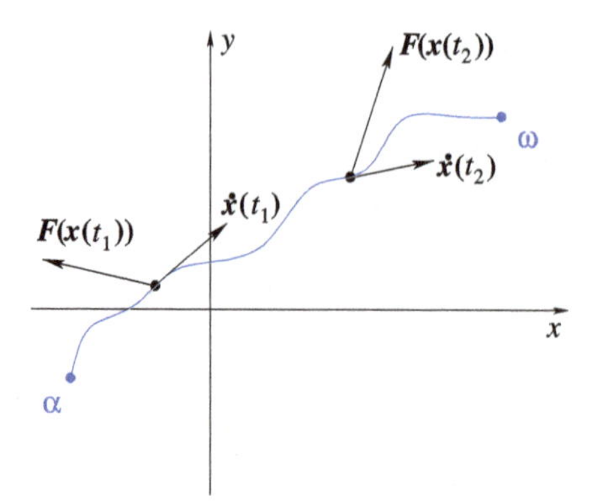

Figure 6.1: The line integral of the vector field $\boldsymbol{F}$ from α to ω along a curve parameterized by the vector function $\boldsymbol{x}$. The derivative $\dot{\boldsymbol{x}}$ is tangent to the curve, pointing in the direction from α to ω. At t_2, the component of $\boldsymbol{F}$ in the direction of $\dot{\boldsymbol{x}}$ is positive, while at t_1, this component is negative.

Proposition 6. *Suppose that two vector functions $\mathbf{x}_1 : [a_1, b_1] \to \mathbb{R}^n$ and $\mathbf{x}_2 : [a_2, b_2] \to \mathbb{R}^n$ both smoothly trace out the same curve C starting at α and ending at ω. For any vector field $\mathbf{F}$ defined on an open set containing C,*

$$\int_C \mathbf{F}(\mathbf{x}) \cdot d\mathbf{x} = \int_{a_1}^{b_1} \mathbf{F}(\mathbf{x}_1(t)) \cdot \dot{\mathbf{x}}_1(t)\, dt = \int_{a_2}^{b_2} \mathbf{F}(\mathbf{x}_2(t)) \cdot \dot{\mathbf{x}}_2(t)\, dt$$

provided that the integrals on the right both exist and are finite. In other words, the value of a line integral depends on the vector field $\mathbf{F}$ and the curve C, but not how that curve is parameterized.

The next proposition allows us to break curves into several pieces, or to reverse the direction of integration.

Proposition 7. *Suppose that C_1 is a smooth curve starting at $\alpha \in \mathbb{R}^n$ and ending at $y \in \mathbb{R}^n$, while C_2 is a smooth curve starting at y and ending at $\omega \in \mathbb{R}^n$. Let $C_1 + C_2$ be the curve starting at α and ending at ω formed by first moving along C_1 and then moving along C_2. Also if C is any smooth curve in $\mathbb{R}^n$, then $-C$ is the same curve, but traversed in the opposite direction. Then*

$$\int_{C_1+C_2} \mathbf{F}(\mathbf{x}) \cdot d\mathbf{x} = \int_{C_1} \mathbf{F}(\mathbf{x}) \cdot d\mathbf{x} = \int_{C_2} \mathbf{F}(\mathbf{x}) \cdot d\mathbf{x}$$

and

$$\int_{-C} \mathbf{F}(\mathbf{x}) \cdot d\mathbf{x} = -\int_{C} \mathbf{F}(\mathbf{x}) \cdot d\mathbf{x}$$

provided that all the integrals exist and are finite.

The proof of the first proposition above is just a change of variables; the proof of the second is left as an exercise (see Exercise 6.2 below).

Remark. Proposition 7 implies that line integrals over piecewise smooth curves are just the sum of line integrals over their smooth sections.

6.1.1 Direct evaluation of line integrals

We begin by directly evaluating several line integrals using the above definition. Later we will see that in some cases, the evaluation of line integrals can be simplified.

Example 6.1. Given a vector field $\boldsymbol{F}(x,y,z) = \langle z,x,y\rangle$ and a smooth curve C parameterized by

$$\begin{cases} x = t^2 \\ y = t \\ z = t^3 \end{cases}$$

for $t \in [0,1]$, please compute

$$\int_C \boldsymbol{F}(\mathbf{x}) \cdot d\mathbf{x}\,.$$

Answer. Straight from the definition:

$$\int_C \boldsymbol{F}(\mathbf{x}) \cdot d\mathbf{x} = \int_0^1 \langle z,x,y\rangle \cdot \left\langle \frac{dx}{dt}, \frac{dy}{dt}, \frac{dz}{dt}\right\rangle\, dt = \int_0^1 \langle t^3,t^2,t\rangle \cdot \langle 2t,1,3t^2\rangle\, dt = \frac{89}{60}$$

It might seem that Proposition 6 implies that it does not matter how a curve is parameterized when a line integral is computed—this is not quite true. While the value of the line integral will be the same for any smooth parameterization, computing the integral may be easier or harder depending on the choice of parameterization; one should always try to choose an easier approach.

Example 6.2. Please evaluate the line integral

$$\int_C \boldsymbol{f}(\mathbf{x}) \cdot d\mathbf{x}$$

where C is the semicircle $x^2+y^2 = 1, y \geq 0$, starting at $\alpha = (1,0)$ and ending at $\omega = (-1,0)$ (see Figure 6.2), and $\boldsymbol{f}(x,y) = \langle -y,x\rangle$.

Answer. At first glance, one might think that setting $x = -t$, $y = \sqrt{1-t^2}$ for $t \in [-1,1]$ would be an effective parameterization. But this approach requires that one carefully simplify and integrate an expression involving a square root. A much easier overall

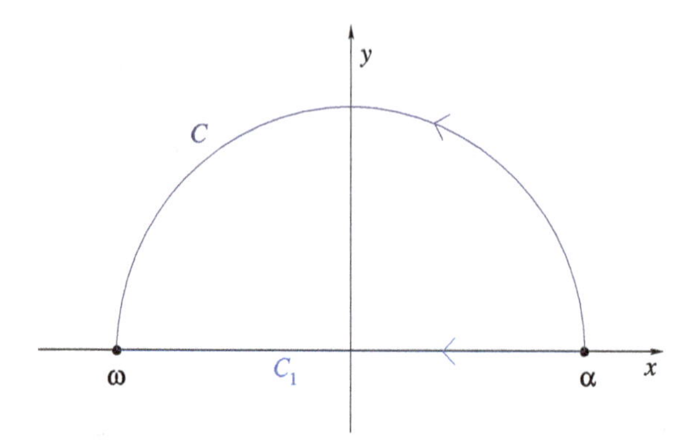

Figure 6.2: The curves C (semicircle in dark blue) and C_1 (line segment in light blue) from α (1, 0) to ω (1, 0). The arrows indicate the direction of integration.

approach is to use trig functions: let $x = \cos t$ and $y = \sin t$. Then

$$\int_C \boldsymbol{f}(\boldsymbol{x}) \cdot d\boldsymbol{x} = \int_0^{\pi} \langle -\sin t, \cos t\rangle \cdot \langle -\sin t, \cos t\rangle\, dt = \int_0^{\pi} (\sin^2 t + \cos^2 t)\, dt = \pi\,.$$

Proposition 6 also does not say that the value of a line integral in general depends only on the endpoints; in general if one changes the integration path (the curve), one changes the integral value.

Example 6.3. Please evaluate the line integral

$$\int_{C_1} \boldsymbol{f}(\boldsymbol{x}) \cdot d\boldsymbol{x}$$

where C_1 is the line segment $y = 0$, $-1 \le x \le 1$, starting at $\alpha = (1, 0)$ and ending at $\omega = (-1, 0)$ (see Figure 6.2), and $\boldsymbol{f}(x, y) = \langle -y, x\rangle$.

Answer. Notice that these are the same endpoints as in the previous example, but not the same curve. In this case, one can simply take $y = 0$, $x = -t$ and integrate with respect to t from -1 to 1:

$$\int_{C_1} \boldsymbol{f}(\boldsymbol{x}) \cdot d\boldsymbol{x} = \int_{-1}^{1} \langle 0, (-t)\rangle \cdot \langle -1, 0\rangle\, dt = -\int_{-1}^{1} 0\, dt = 0\,.$$

6.1.2 Path dependence; path independence

In the previous section, line integrals were computed along curves that start at a point α and end at a point a point ω. One might wonder, given a certain α and ω, does it matter which curve (path) is used to connect them? In general, as we saw in Example 6.2 and Example 6.3, the answer is "yes": different integration paths lead to different values for the integral. But there is an important class of vector fields whose line integrals depend only on the starting and ending points α and ω, and *not* on the integration path or curve C that connects them.

Definition. A vector field $\boldsymbol{F}$ is *conservative* on its open domain if and only if, given any two points α and ω in the domain, the value of the line integral from α to ω is the same, independent of which path or curve C through the domain is chosen to move from α to ω:

$$\int_C \boldsymbol{f}(\mathbf{x}) \cdot d\mathbf{x} = \int_\alpha^\omega \boldsymbol{f}(\mathbf{x}) \cdot d\mathbf{x}$$

for all curves C starting at α and ending at ω.

Remark. There is a technical point here that is worth mentioning: Saying that a vector field "is conservative on its open domain" implicitly requires that this domain is pathwise connected since the definition of conservative requires that there is at least one curve between any two given points in the domain.

Conservative vector fields are among the most important in nature. Electrical fields are conservative, and gravitational fields are conservative when friction forces can be neglected. The applications of these mathematical results are not discussed here, but they should be the central part of any good treatment of basic physics.

One immediate consequence of a vector field being conservative is that a line integral around any closed curve (a curve that begins and ends at the same point, so $\alpha = \omega$) is zero.

Theorem 18. *A vector field* $\boldsymbol{F}$ *is conservative on its open domain* $D \subset \mathbb{R}^n$ *if and only if the line integral over any closed curve* $\mathcal{O} \subset D$ *is zero:*

$$\oint_{\mathcal{O}} \boldsymbol{F}(\mathbf{x}) \cdot d\mathbf{x} = 0$$

Proof. First, assume that the line integral over any closed curve $\mathcal{O}$ is zero, and suppose that C_1 and C_2 are any two curves lying in the domain $D \subset \mathbb{R}^n$ that each begin at the point α and end at the point ω. Then $C_1 - C_2 = C_1 + (-C_2)$ is a closed curve, and if $\boldsymbol{F}$ is conservative, then

$$\int_{C_1 - C_2} \boldsymbol{F}(\mathbf{x}) \cdot d\mathbf{x} = \int_{C_1} \boldsymbol{F}(\mathbf{x}) \cdot d\mathbf{x} - \int_{C_2} \boldsymbol{F}(\mathbf{x}) \cdot d\mathbf{x} = 0$$

implying of course that

$$\int_{C_1} \boldsymbol{F}(\mathbf{x}) \cdot d\mathbf{x} = \int_{C_2} \boldsymbol{F}(\mathbf{x}) \cdot d\mathbf{x}\,.$$

The proof for the opposite direction is perhaps most easily done by considering the contrapositive: assume that there is a closed curve $\mathcal{O}$ over which the line integral is *not* zero. Pick any two distinct points on $\mathcal{O}$ as α and ω; let C_1 be one portion of $\mathcal{O}$ moving from α to ω, and let C_2 be the other portion. Then since

$$\oint_{\mathcal{O}} \boldsymbol{F}(\boldsymbol{x}) \cdot d\boldsymbol{x} = \int_{C_1 - C_2} \boldsymbol{F}(\boldsymbol{x}) \cdot d\boldsymbol{x} = \int_{C_1} \boldsymbol{F}(\boldsymbol{x}) \cdot d\boldsymbol{x} - \int_{C_2} \boldsymbol{F}(\boldsymbol{x}) \cdot d\boldsymbol{x} \neq 0$$

implying that

$$\int_{C_1} \boldsymbol{F}(\boldsymbol{x}) \cdot d\boldsymbol{x} \neq \int_{C_2} \boldsymbol{F}(\boldsymbol{x}) \cdot d\boldsymbol{x} .$$

Thus there is no path independence in this case. □

It would seem clear that if a vector field is conservative, then computing line integrals of this vector field should be simpler, but the obvious questions are (1) How does one recognize that a vector field is conservative? and (2) How does one take advantage of the fact that the vector field is conservative? Happily, these questions both have at their root the same answer: find a potential function.

Definition. A differentiable scalar function $\varphi : D \subset \mathbb{R}^n \to \mathbb{R}$ is a *potential function* for a vector field $\boldsymbol{F}$ if and only if $\boldsymbol{F}(\boldsymbol{x}) = \nabla\varphi(\boldsymbol{x})$ for all $\boldsymbol{x} \in D \subset \mathbb{R}^n$.[1]

The really important thing is the connection between the previous two definitions.

Theorem 19. *A vector field $\boldsymbol{F}$ is conservative on its open domain $D \subset \mathbb{R}^n$ if and only if there is a potential function φ defined on the same domain such that $\boldsymbol{F}(\boldsymbol{x}) = \nabla\varphi(\boldsymbol{x})$ for all $\boldsymbol{x} \in D$.*

Proof. Only the backward direction of this proof is discussed here; the forward direction is discussed in Exercise 6.7. Suppose that a potential function φ exists with $\boldsymbol{F} = \nabla\varphi$. Suppose that the vector function $\boldsymbol{x}$ smoothly traces out the curve C beginning at $t = t_o$ and ending at $t = t_1$. Then using the chain rule and a changing variables, one can reduce the line integral to a single-variable integral:

$$\begin{aligned}\int_C \boldsymbol{F}(\boldsymbol{x}) \cdot d\boldsymbol{x} &= \int_C \nabla\varphi(\boldsymbol{x}) \cdot \frac{d\boldsymbol{x}}{dt}\, dt = \int_{t_o}^{t_1} \left(\frac{\partial\varphi}{\partial x_1}\frac{dx_1}{dt} + \cdots + \frac{\partial\varphi}{\partial x_n}\frac{dx_n}{dt} \right) dt \\ &= \int_{t_o}^{t_1} \frac{d}{dt}(\varphi(\boldsymbol{x}(t)))\, dt = \varphi(\boldsymbol{x}(t))\Big|_{t_o}^{t_1} = \varphi(\omega) - \varphi(\alpha)\end{aligned}$$

where t_o corresponds to α (the beginning of the curve) and t_1, to ω (the end of the curve). □

1 In physics, particularly in electromagnetism, there may be a negative sign: $\boldsymbol{E} = -\nabla V$ where $\boldsymbol{E}$ is the electric (vector) field and V is the electrical potential (function). This sign is just a matter of convenience or inconvenience, depending on ones point of view, but the general definition of potential function as given here does not include a negative sign.

Example 6.4. Please evaluate the line integral

$$\int_C \boldsymbol{f}(\boldsymbol{x}) \cdot d\boldsymbol{x}$$

where C is *any* curve starting at $\alpha = (1,1)$, ending at $\omega = (3,5)$, and $\boldsymbol{f}(x,y) = \langle 2x + y, x + 2y \rangle$.

Answer. In this case, if there is going to be a single value for the integral, it must be the same no matter which curve between α to ω is chosen. This suggests that we should look for a potential function. The method for finding this potential function can be called *partial integration* because it is essentially the reverse of partial differentiation. If the potential function φ exists, then from its definition, we know that the components of $\boldsymbol{f}$ must equal the partial derivatives of φ:

$$f_1 = \frac{\partial\varphi}{\partial x} = 2x + y\,, \quad f_2 = \frac{\partial\varphi}{\partial y} = x + 2y\,.$$

Thus we can find how φ depends on x by integrating the first of these two equations with respect to x (and treating y as a constant), and then we can differentiate with respect to y:

$$\varphi(x,y) = x^2 + xy + c(y) \quad \Longrightarrow \quad \frac{\partial\varphi}{\partial y} = x + c'(y)$$

where c is the integration constant which here may depend on y. The question now is can the two expressions for $\partial\varphi/\partial y$ be reconciled? In the present example, the answer is yes, provided that $c'(y) = 2y$, and thus $c(y) = y^2$. One may add any constant, but it is usually convenient to take this constant to be zero. Hence in this example, $\varphi(x,y) = x^2 + xy + y^2$.

Finally, to compute the actual line integral, one needs only to compute the difference between the potential function evaluated at the two points:

$$\int_C \boldsymbol{f}(\boldsymbol{x}) \cdot d\boldsymbol{x} = \int_{(1,1)}^{(3,5)} \boldsymbol{f}(\boldsymbol{x}) \cdot d\boldsymbol{x} = \varphi(x,y)\Big|_{(1,1)}^{(3,5)} = x^2 + xy + y^2\Big|_{(1,1)}^{(3,5)} = 49 - 3 = 46$$

What happens if one attempts the partial integration process discussed above on a vector field that does not have any potential function? The next example addresses this situation.

Example 6.5. Show that no potential function exists for the vector field $\boldsymbol{f}(x,y) = \langle 2xy, 2xy \rangle$.

Answer. As in the previous example, let us try to find a potential function φ. Again we know that the components of $\boldsymbol{f}$ must equal the partial derivatives of φ:

$$f_1 = \frac{\partial\varphi}{\partial x} = 2xy\,, \quad f_2 = \frac{\partial\varphi}{\partial y} = 2xy$$

Integrating the first of these equations with respect to x, one finds that

$$\varphi(x,y) = x^2y + c(y) \quad \Longrightarrow \quad \frac{\partial\varphi}{\partial y} = x^2 + c'(y)\,.$$

But from the second component of $\boldsymbol{f}$, one then has

$$x^2 + c'(y) = 2xy \quad \Longrightarrow \quad c'(y) = 2xy - x^2\,.$$

This final equation is a contradiction because $c(y)$ must be a function of y alone—it must be constant with respect to x. Here, this requirement cannot be satisfied, hence no potential function φ is possible.

6.1.3 Flow *crossing* a curve

Up to this point, this section has dealt only with integration of the component of a vector field *along* a curve, and hence flow *along* a curve; now we consider integration that deals with flow *across* a curve. Line integrals representing flow along a curve are the primary form of line integrals, but there are cases where this second form is important. In particular, this second type of line integral is used to describe flow into and out of regions. This type of line integral generally occurs only in $\mathbb{R}^2$ (the x,y-plane).

Definition. Let C be a smooth curve in $\mathbb{R}^2$ traced out by a differentiable vector function $\boldsymbol{x} : [a,b] \to \mathbb{R}^2$ with $\boldsymbol{x}(a)$ being the starting point α for C and $\boldsymbol{x}(b)$ being the ending point ω. Let $\boldsymbol{F} : \mathbb{R}^2 \to \mathbb{R}^2$ be a vector field, and let $\boldsymbol{n}$ be a unit normal vector to the curve C. Then the *line integral crossing C in the direction of $\boldsymbol{n}$* is

$$\int_C \boldsymbol{F}(\boldsymbol{x}) \cdot \boldsymbol{n}\, ds := \int_a^b \boldsymbol{F}(\boldsymbol{x}(t)) \cdot \boldsymbol{n}(s(t))\,\frac{ds}{dt}\,dt \equiv \int_a^b \boldsymbol{F}(\boldsymbol{x}(t)) \cdot \boldsymbol{n}(s(t))\,\dot{s}(t)\,dt$$

where s is arc length along the curve C starting from α provided that the integral on the right exists and is finite.

Remark. If C is a simple closed curve, than $\boldsymbol{n}$ is normally taken to be the *outward* unit normal vector. For other curves, one must simply be careful to notice which unit normal vector is being used.

Example 6.6. Please evaluate the line integral

$$\int_C \boldsymbol{f}(\boldsymbol{x}) \cdot \boldsymbol{n}\, ds$$

where C is the circle $x^2+y^2 = 4$, $\boldsymbol{n}$ is the outward unit normal vector, and $\boldsymbol{f}(x,y) = \langle y,x\rangle$.

Answer. As before, if one is integrating around a circle, one should likely use polar coordinates; let $x = 2\cos t$ and $y = 2\sin t$. Then $\dot{s}(t) = |\boldsymbol{v}(t)| = \sqrt{(-2\sin t)^2 + (2\cos t)^2} = 2$. Because C is a circle, the outward unit normal vector is the radial vector: $\boldsymbol{n} = \boldsymbol{x}(t)/|\boldsymbol{x}(t)|$. In this case, $\boldsymbol{n} = \langle \cos t, \sin t\rangle$. This

$$\int_C \boldsymbol{f}(\boldsymbol{x}) \cdot \boldsymbol{n}\, ds = \int_0^{2\pi} \langle \sin t, \cos t\rangle \cdot \langle \cos t, \sin t\rangle\, (2)\, dt = 4 \int_0^{2\pi} \sin t \cos t\, dt = 0\,.$$

Notice that in this example the flow across the circle is generally not zero at any given point on the circle, but the integral above shows that the net flow is zero.

It is, of course, possible to define still other types of line integrals, but those would be unusual; the two forms defined here (along the curve, and across the curve) are by far the most common types to appear in mathematics, science, and engineering.

6.2 Surface integrals: integration over surfaces in $\mathbb{R}^3$

As it is possible to integrate along curves that are not intervals on the x-axis, it is also possible to integrate over surfaces that are not simply portions of the x, y-plane. Unlike the line integral case, however, the primary component of integration for the integrand for surface integrals is the component *normal* to the surface.

Definition. Let S be a smooth surface in $\mathbb{R}^3$ that is described by the equation $z = s(x, y)$ where $s : D \subset \mathbb{R}^2 \to \mathbb{R}$ is a continuously differentiable function defined on some domain D in the x, y-plane. Let $\boldsymbol{F} : \mathbb{R}^3 \to \mathbb{R}^3$ be a vector field, and let $\boldsymbol{n}$ be a unit normal vector to the surface S. Then the *surface integral* crossing S in the direction of $\boldsymbol{n}$ is

$$\iint_S \boldsymbol{F}(\boldsymbol{x}) \cdot \boldsymbol{n}\, dS := \iint_D (\boldsymbol{F}(x, y, s(x, y)) \cdot \boldsymbol{n}) \sqrt{(\partial s/\partial x)^2 + (\partial s/\partial y)^2 + 1}\; dA$$

provided that the integral on the right exists and is finite.[2] The components of this surface integral are shown in Figure 6.3.

Remarks.

1. The square-root factor in the definition of the surface integral comes from the tilt of the surface S relative to the x, y-plane. It is 1 when S is horizontal, greater than 1 when S is tilted, and will vary with x and/or y unless S is a plane.

---

2 This is not the most general definition for surface integrals, since it requires that the surface lie above some domain in the x, y-plane; basic closed surfaces like spheres cannot be fully described in this way. Still this definition can be used to compute and discuss all of the surface integrals that we are interested in here, including integrals over spheres.

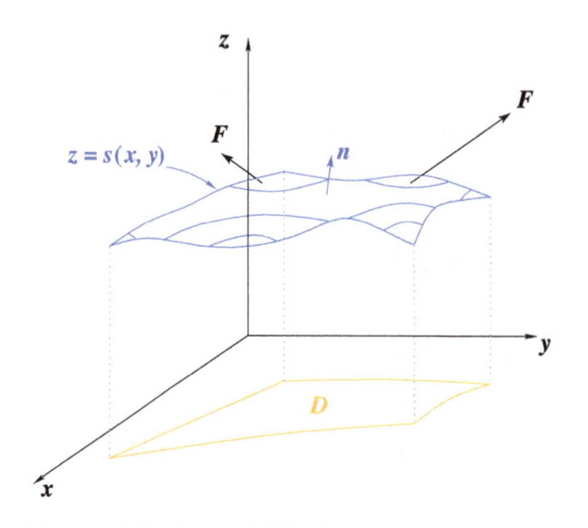

Figure 6.3: The integral of the vector field $\boldsymbol{F}$ over the surface $z = s(x, y)$ (in blue) above an integration domain D (in gold).

2. In this integral, dA will $dx\,dy = dy\,dx$ in rectangular coordinates, or $r\,dr\,d\theta$ in polar coordinates.

The next two examples show the direct evaluation of two surface integrals using this definition.

Example 6.7. Please evaluate the surface integral

$$\iint_{\Pi} \boldsymbol{F} \cdot \boldsymbol{n}\, dS$$

where $\boldsymbol{F}(x, y, z) = \langle z, x, y\rangle$, Π is the portion of the plane $2x + 3y + z = 1$ lying in the first octant, and $\boldsymbol{n}$ is the upper unit normal vector for this plane.

Answer. Given this surface Π, the corresponding integration domain D is the triangular region bounded by the x-axis, the y-axis, and the line $y = (1 - 2x)/3$. Also since Π is a plane, one can read off the normal vector $\boldsymbol{N} = \langle 2, 3, 1\rangle$, implying that the upper unit normal vector is $\boldsymbol{n} = \langle 2, 3, 1\rangle/\sqrt{14}$. The integral is then

$$\begin{aligned}\iint_{\Pi} \boldsymbol{F} \cdot \boldsymbol{n}\, dS &= \int_0^{1/2}\int_0^{(1-2x)/3} \langle 1 - 2x - 3y, x, y\rangle \cdot \frac{\langle 2, 3, 1\rangle}{\sqrt{14}} \sqrt{(-2)^2 + (-3)^2 + 1}\, dy\, dx \\ &= \int_0^{1/2}\int_0^{(1-2x)/3} (2 - x - 5y)\, dy\, dx = \int_0^{1/2} \left(2y - xy - \frac{5}{2}y^2\Big|_0^{(1-2x)/3}\right. dx \\ &= \frac{1}{9}\int_0^{1/2} \left(\frac{7}{2} - 5x - 4x^2\right) dx = \frac{23}{216}.\end{aligned}$$

Example 6.8. Please evaluate the surface integral

$$\iint_S \boldsymbol{F} \cdot \boldsymbol{n}\, dS$$

where $\boldsymbol{F}(\rho, \phi, \theta) = \boldsymbol{\rho}$, $\boldsymbol{\rho}$ being the vector from the origin to a point in three-space, S is the sphere of radius 2 centered at the origin, and $\boldsymbol{n}$ is the outward unit normal vector for this sphere.

Answer. Technically, this surface does not fit the definition for a surface integral given above because the *entire* surface cannot be expressed as $z = s(x, y)$, but this issue can be dealt with by dividing the sphere into two parts: the upper hemisphere and the lower hemisphere. First, consider the upper hemisphere; the corresponding integration domain D is the circular disk in the x, y-plane centered at the origin with radius 2. The upper hemisphere H is $z = s(x, y) = \sqrt{4 - x^2 - y^2}$, and thus

$$\frac{\partial s}{\partial x} = \frac{-x}{\sqrt{4 - x^2 - y^2}}, \qquad \frac{\partial s}{\partial y} = \frac{-y}{\sqrt{4 - x^2 - y^2}}.$$

For any point on the sphere, $\boldsymbol{\rho}$ is the vector from the origin to that point; because of the symmetry of the sphere, the outward unit normal vector is $\boldsymbol{n} = \boldsymbol{\rho}/\rho$. The integral for the upper hemisphere is

$$\begin{aligned}
\iint_H \boldsymbol{F} \cdot \boldsymbol{n}\, dS &= \int_0^{2\pi}\int_0^2 (\boldsymbol{\rho} \cdot \boldsymbol{\rho}/\rho)\sqrt{\left(\frac{-x}{\sqrt{4 - x^2 - y^2}}\right)^2 + \left(\frac{-y}{\sqrt{4 - x^2 - y^2}}\right)^2 + 1}\; r\, dr\, d\theta \\
&= \int_0^{2\pi}\int_0^2 (\rho)\left(\frac{2}{\sqrt{4 - r^2}}\right) r\, dr\, d\theta \\
&= 8\pi \int_0^2 \frac{r\, dr}{\sqrt{4 - r^2}} = 16\pi
\end{aligned}$$

since $\rho = 2$ on the surface. The integral over the lower hemisphere is the same because of symmetry; for the entire sphere S then,

$$\iint_S \boldsymbol{F} \cdot \boldsymbol{n}\, dS = 32\pi .$$

6.3 Differential operators

6.3.1 Definitions

There are several "collections" of partial derivatives that arise naturally in mathematics and science; three of these have already come up in our discussion: the gradient vector and the Jacobian and Hessian matrices. We now introduce three more: the divergence, the curl, and the Laplacian.[3] As the names suggest, the divergence measures or describes the tendency of a vector field to diverge from a point, while curl measures or describes the tendency of a vector field to circulate around a point (see below for details).

Definition. Suppose that $\boldsymbol{u} : D \subset \mathbb{R}^3 \to \mathbb{R}^3$ is a vector field. The *divergence* of this vector field $\boldsymbol{u}$ is defined as

$$\operatorname{div} \boldsymbol{u} \equiv \nabla \cdot \boldsymbol{u} := \frac{\partial u_1}{\partial x_1} + \frac{\partial u_2}{\partial x_2} + \frac{\partial u_3}{\partial x_3}$$

provided that these partial derivatives exist.

Definition. Suppose again that $\boldsymbol{u} : D \subset \mathbb{R}^3 \to \mathbb{R}^3$ is a vector field. The *curl* of this vector field $\boldsymbol{u}$ is defined as

$$\operatorname{curl} \boldsymbol{u} \equiv \nabla \times \boldsymbol{u} := \left\langle \frac{\partial u_3}{\partial x_2} - \frac{\partial u_2}{\partial x_3}, \frac{\partial u_1}{\partial x_3} - \frac{\partial u_3}{\partial x_1}, \frac{\partial u_2}{\partial x_1} - \frac{\partial u_1}{\partial x_2} \right\rangle \equiv \begin{vmatrix} \mathbf{i} & \mathbf{j} & \mathbf{k} \\ \frac{\partial}{\partial x_1} & \frac{\partial}{\partial x_2} & \frac{\partial}{\partial x_3} \\ u_1 & u_2 & u_3 \end{vmatrix}$$

provided that these partial derivatives exist.

Definition. Suppose that $v : D \subset \mathbb{R}^3 \to \mathbb{R}$ is a scalar multivariable function. The *Laplacian* of this scalar function v is defined as

$$\Delta v := \frac{\partial^2 v}{\partial x^2} + \frac{\partial^2 v}{\partial y^2} + \frac{\partial^2 v}{\partial z^2}$$

provided that all of these partial derivatives exist.

Remarks.
1. For any $n \in \mathbb{Z}^+$, the definition of the divergence can be given for a vector field on $\mathbb{R}^n$, and the Laplacian can be defined for a scalar function on $\mathbb{R}^n$, but as with the cross product, the curl requires that $n = 3$, that is, three-dimensional space.
2. As has been the case throughout this text, the numbering of variables (x_1, x_2, x_3) is interchanged with the use of traditional letters (x, y, z).

6.3.2 Why is div($\boldsymbol{u}$) actually divergence?

The definition for divergence given above is simple and standard, but at first glance, it may not be at all clear why div($\boldsymbol{u}$) should represent divergence. What follows is both a justification for this definition, and in essence, the basis for a proof for the divergence theorem presented in the following section. This presentation is for $n = 2$ (two-

3 Named in honor of Pierre-Simon Laplace (1749–1827), a French mathematician, scientist, and engineer from roughly two generations after Leibniz and Newton.

dimensional space), but the same sort of argument works for n-dimensional space. Also a similar argument justifies the definition for curl (see Exercise 6.13).

Theorem 20. *Suppose that a vector field* $\boldsymbol{u}$ *is continuously differentiable:* $\boldsymbol{u} \in C^1$ *on some open set containing a point* $(x, y) \in \mathbb{R}^2$*. Then*

$$\operatorname{div}(\boldsymbol{u})(x,y) = \lim_{\epsilon \to 0} \frac{1}{|A|} \oint_{\partial A} \boldsymbol{u} \cdot \boldsymbol{n}\, ds \tag{6.1}$$

where A *is a square centered at* (x, y) *having edge length* $\epsilon > 0$*, the boundary of* A *is* ∂A*, and* $|A|$ *is the area of* A *(see Figure 6.4).*

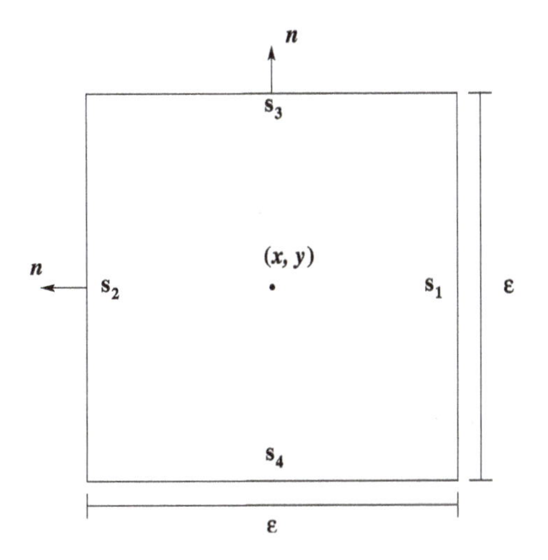

Figure 6.4: Square centered at (x, y) with edge length ϵ. The vertical edges are s_1 and s_2; the horizontal edges are s_3 and s_4. The outward unit normal vectors on each edge are $\boldsymbol{n}$.

Remark. Notice that the line integral in this theorem is of the second form and that the center (x, y) is the only point in every square no matter how small ϵ is. Theorem 20 states that the divergence of a vector field $\boldsymbol{u}$ from (x, y) is the average value of the outward flow of $\boldsymbol{u}$ crossing the boundary of smaller and smaller squares as $\epsilon \to 0$. Thus this quantity is the tendency for the vector field $\boldsymbol{u}$ to diverge from the point (x, y).

Proof. The line integral (6.1) can be computed explicitly by breaking it into four edge segments as shown in Figure 6.4. Consider first the two vertical segments, s_1 and s_2. For s_1,

$$\int_{s_1} \boldsymbol{u} \cdot \boldsymbol{n}\, ds = \int_{y-\epsilon/2}^{y+\epsilon/2} \boldsymbol{u}(x+\epsilon/2, \eta) \cdot \langle 1, 0 \rangle \, d\eta = \int_{y-\epsilon/2}^{y+\epsilon/2} u_1(x+\epsilon/2, \eta)\, d\eta$$

where u_1 is the first component of $\boldsymbol{u}$. Using a Taylor expansion, one can write

$$u_1(x+\epsilon/2, \eta) = u_1(x+\epsilon/2, y) + \frac{\partial u_1}{\partial y}(x+\epsilon/2, y)(\eta - y) + O(\epsilon^2)$$

where $O(\epsilon^2)$ represent the remaining terms in the expansion that are of the order of ϵ^2 or higher. Hence

$$\frac{1}{|A|}\int_{s_1} \boldsymbol{u}\cdot\boldsymbol{n}\,ds = \frac{u_1(x+\epsilon/2,y)}{\epsilon} + \frac{\partial u_1}{\partial y}(x+\epsilon/2,y) + O(\epsilon) \tag{6.2}$$

since $|A| = \epsilon^2$, and $u_1(x+\epsilon/2,y)$ and $(\partial u_1/\partial y)(x+\epsilon/2,y)$ are constants with respect to this integration. The same sort of computation holds for s_2, except that now, the integral is from $y+\epsilon/2$ to $y-\epsilon/2$ (because one moves counterclockwise around ∂A, and $\boldsymbol{n} = \langle -1, 0\rangle$). As a result,

$$\int_{s_2} \boldsymbol{u}\cdot\boldsymbol{n}\,ds = \int_{y+\epsilon/2}^{y-\epsilon/2} \boldsymbol{u}(x-\epsilon/2,\eta)\cdot\langle -1,0\rangle\,(-d\eta) = -\int_{y-\epsilon/2}^{y+\epsilon/2} u_1(x-\epsilon/2,\eta)\,d\eta$$

which implies that

$$\frac{1}{|A|}\int_{s_2} \boldsymbol{u}\cdot\boldsymbol{n}\,ds = -\frac{u_1(x-\epsilon/2,y)}{\epsilon} - \frac{\partial u_1}{\partial y}(x-\epsilon/2,y) + O(\epsilon) \tag{6.3}$$

Combining (6.2) and (6.3), one finds that

$$\lim_{\epsilon\to 0}\frac{1}{|A|}\int_{s_1+s_2} \boldsymbol{u}\cdot\boldsymbol{n}\,dA = \lim_{\epsilon\to 0}\frac{u_1(x+\epsilon/2,y) - u_1(x-\epsilon/2,y)}{\epsilon} = \frac{\partial u_1}{\partial x}(x,y)$$

because the continuity of the partial derivatives implies that the $\partial u_1/\partial y$ terms in (6.2) and (6.3) cancel.

Now consider the two horizontal edge segments: s_3 and s_4. The same argument as before leads to

$$\lim_{\epsilon\to 0}\frac{1}{|A|}\int_{s_3+s_4} \boldsymbol{u}\cdot\boldsymbol{n}\,dA = \lim_{\epsilon\to 0}\frac{u_2(x,y+\epsilon/2) - u_2(x,y-\epsilon/2)}{\epsilon} = \frac{\partial u_2}{\partial y}(x,y)\,,$$

and adding the four edge segments together yields that

$$\lim_{\epsilon\to 0}\frac{1}{|A|}\oint_{\partial A} \boldsymbol{u}\cdot\boldsymbol{n}\,dA = \frac{\partial u_1}{\partial x}(x,y) + \frac{\partial u_2}{\partial y}(x,y) = \operatorname{div}(\boldsymbol{u})(x,y)\,.$$

□

Remark. The argument above uses rectangular coordinates; one might ask whether or not this result really requires rectangular coordinates. The answer is that it does not; see Exercise 6.14.

6.4 The theorems of Gauss, Green, and Stokes

In many ways, this is the main section, not only of this chapter, but of the entire text. Here we discuss the work from the early 19th century of a number of mathematicians, including Gauss (or Gauß), Green, and Stokes. Of particular note is the contribution of George Green who proved most of these results while working as a miller. Indeed in many ways, calculus began with the work of Leibniz and Newton and was complete with the work of Green. Work in the second half of the 19th century and later is better thought of as analysis rather than calculus.

6.4.1 The divergence theorem

Theorem 21 (Divergence, Gauss[4]). *Suppose that $\Omega \subset \mathbb{R}^n$ is a bounded, open, connected domain, and suppose that $\partial\Omega$ (the boundary of Ω) is piecewise smooth. Suppose that $\boldsymbol{u} \in C^1(\overline{\Omega})$, meaning that $\boldsymbol{u}$ is continuously differentiable on the interior of Ω and continuous up to its boundary. Then*

$$\int_{\Omega} \operatorname{div} \boldsymbol{u}\, dV = \oint_{\partial\Omega} \boldsymbol{u} \cdot \boldsymbol{n}\, dS$$

where $\boldsymbol{n}$ is the outer *unit normal vector to $\partial\Omega$.*

Proof. Consider a domain $\Omega \subset \mathbb{R}^2$ whose boundary $\partial\Omega$ is a closed curve (see Figure 6.5). Given $\epsilon > 0$, partition Ω into sufficiently small squares ΔA_{ij} with areas $|\Delta A_{ij}| = \epsilon^2$ where

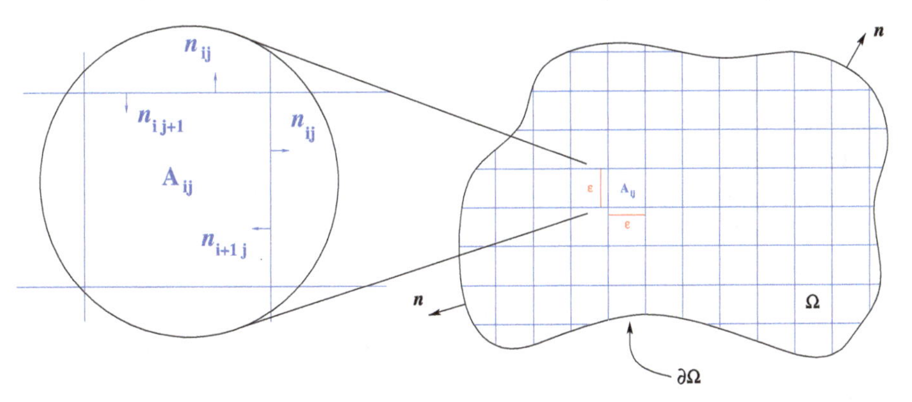

Figure 6.5: Domain for integration Ω for the divergence theorem. The partition is in blue, and each partition square A_{ij} has area ϵ^2; partition elements near the boundary have smaller area.

4 This theorem is widely attributed to Gauss (1777–1855) in both mathematics and physics. A version was known earlier to Lagrange (1736–1813), and independent proofs were given by Ostrogradsky (1801–1862) and Green (1793–1841).

i is the index for the x coordinate (the column in Figure 6.5) and j is the index for the y coordinate (the row in Figure 6.5). Near the boundary, the partition elements are not necessarily square, but their areas are still no greater than ϵ^2. For each of the ΔA_{ij}, let (x_i, y_j) be the center of ΔA_{ij}. From Theorem 20, at each center (x_i, y_j), we can pick a sufficiently small ϵ so that

$$\operatorname{div}(\boldsymbol{u})(x_i, y_j) = \lim_{|\Delta A_{ij}| \to 0} \frac{1}{|\Delta A_{ij}|} \oint_{\partial \Delta A_{ij}} \boldsymbol{u} \cdot \boldsymbol{n}\, ds = \frac{1}{\epsilon^2} \oint_{\partial \Delta A_{ij}} \boldsymbol{u} \cdot \boldsymbol{n}\, ds + O(\epsilon)$$

where the error term $O(\epsilon)$ goes to zero as $\epsilon \to 0$.

Now, multiplying by $|\Delta A_{ij}| = \epsilon^2$ and summing over all of the partition elements, one finds that

$$\sum_{i,j} \operatorname{div}(\boldsymbol{u})(x_i, y_j) |\Delta A_{ij}| = \sum_{i,j} \left(\oint_{\partial \Delta A_{ij}} \boldsymbol{u} \cdot \boldsymbol{n}\, ds + O(\epsilon^3) \right) = \oint_{\partial \Omega} \boldsymbol{u} \cdot \boldsymbol{n}\, ds + O(\epsilon)$$

where the second equality is due to the cancellation of line integrals on the boundaries of adjacent partition elements (see Figure 6.5). For example, the flow rightward out of ΔA_{ij} is equal and opposite the flow leftward out of $\Delta A_{i+1,j}$ because $\boldsymbol{n}_{ij} = -\boldsymbol{n}_{i+1j}$ along their common boundary, and the flow upward out of ΔA_{ij} is equal and opposite the flow downward out of $\Delta A_{i,j+1}$. The error term goes from $O(\epsilon^3)$ to $O(\epsilon)$ because there are $O(1/\epsilon^2)$ partition squares.

Finally, taking the limit as $|\Delta A_{ij}| \to 0$, one finds that

$$\iint_{\Omega} \operatorname{div}(\boldsymbol{u}) dA = \lim_{|\Delta A_{ij}| \to 0} \sum_{i,j} \operatorname{div}(\boldsymbol{u})(x_i, y_j) |\Delta A_{ij}| = \oint_{\partial \Omega} \boldsymbol{u} \cdot \boldsymbol{n}\, ds \qquad \square$$

Remarks.

1. The integral on the left in the divergence theorem is a multiple integral with as many integrations as the dimension of the space in which Ω is embedded—here, $\mathbb{R}^n$. The integral on the right has one fewer integrations since it is over the boundary. So if $n = 3$,

$$\iiint_{\Omega} \operatorname{div} \boldsymbol{u}\, dV = \oiint_{\partial \Omega} \boldsymbol{u} \cdot \boldsymbol{n}\, dS,$$

   whereas if $n = 2$,

$$\iint_{\Omega} \operatorname{div} \boldsymbol{u}\, dA = \oint_{\partial \Omega} \boldsymbol{u} \cdot \boldsymbol{n}\, ds.$$

   For $n = 2$, the integral over $\partial \Omega$ is a line integral of the second type.

2. Physically and mathematically, what this theorem is saying is that the total amount that a vector field flows out of or diverges from the inside of a domain Ω is equal to the amount that this vector field crosses out of the domain boundary $\partial\Omega$. Thus this theorem is a statement of conservation of flow for the vector field. If the divergence is negative, then the vector field flows into the domain Ω, and this must equal to the amount that the vector field crosses into the domain boundary $\partial\Omega$. Suppose that $f > 0$ is a given function defined on a domain Ω; if div $\boldsymbol{u} = f > 0$ throughout Ω, then f is the source function for a vector field $\boldsymbol{u}$ and the total flow crossing the boundary out of Ω is

$$\int_{\Omega} f\, dV = \oint_{\partial\Omega} \boldsymbol{u} \cdot \boldsymbol{n}\, dS\,.$$

3. This is not the most general statement of the divergence theorem, for example, the domain of integration need not be connected. Still this version demonstrates all of the essential mathematics of the theorem.

In many cases, the divergence theorem can be used to turn a complicated integral into a much simpler integral. Frequently it is much easier to compute the integral over Ω than to compute the integral over its boundary, as the following example demonstrates.

Example 6.9. Given a vector field $\boldsymbol{F}(x, y, z) = \langle 3x, 2y, z\rangle$, please find the value of the surface integral over the surface of the unit cube $C := \{(x, y, z) \mid 0 < x < 1, 0 < y < 1, 0 < z < 1\}$:

$$\iint_{\partial C} \boldsymbol{F} \cdot \boldsymbol{n}\, dS$$

where $\boldsymbol{n}$ is the outward unit normal vector.

Answer. Computing this integral directly would require that one compute six separate surface integrals. But since div $\boldsymbol{F} = 3 + 2 + 1 = 6$, the divergence theorem implies that

$$\iint_{\partial C} \boldsymbol{F} \cdot \boldsymbol{n}\, dS = \iiint_{C} \operatorname{div} \boldsymbol{F}\, dV = 6 \iiint_{C} dV = 6$$

because of course the volume of the unit cube is 1.

It is also at times possible to use the divergence theorem to compute surface integrals for surfaces that do not by themselves completely bound domains, as the following example shows.

Example 6.10. Given a vector field $\boldsymbol{u}(x, y, z) = \langle -y^2 + z, x^2 + z, -xy^2 + z^2\rangle$, please find the value of the surface integral over the surface of the hemisphere $H := \{(x, y, z) \mid$

$x^2 + y^2 + z^2 = 3,\ z > 1\}$:

$$\iint_H \boldsymbol{u} \cdot \boldsymbol{n}\, dS$$

where $\boldsymbol{n}$ is the upward unit normal vector.

Answer. Computing this integral directly again would be at least somewhat tedious, but taking into account the circular disk $D := \{(x, y, 0) \mid x^2 + y^2 \leq 3\}$ in the x, y-plane, one can then use the divergence theorem to compute the given integral:

$$\iint_H \boldsymbol{u} \cdot \boldsymbol{n}\, dS + \iint_D \boldsymbol{u} \cdot \boldsymbol{n}\, dS = \iiint_\Omega \operatorname{div} \boldsymbol{u}\, dV$$

where Ω is the domain inside the hemisphere so that $\partial\Omega = H \cup D$ and $\boldsymbol{n} = \langle 0, 0, 1 \rangle$. Hence

$$\iint_H \boldsymbol{u} \cdot \boldsymbol{n}\, dS = \iiint_\Omega \operatorname{div} \boldsymbol{u}\, dV - \iint_D \boldsymbol{u} \cdot \boldsymbol{n}\, dS .$$

Now

$$\begin{aligned} \iiint_\Omega \operatorname{div} \boldsymbol{u}\, dV &= \iiint_\Omega 2z\, dV = 2 \int_0^{2\pi} \int_0^{\pi/2} \int_0^{\sqrt{3}} (\rho \cos\phi) \left(\rho^2 \sin\phi\right) d\rho\, d\phi\, d\theta \\ &= 2\,(2\pi) \left(\int_0^1 w\, dw \right) \left(\int_0^{\sqrt{3}} \rho^3\, d\rho \right) = (2\pi)\,(1) \left(\frac{9}{4}\right) = \frac{9\pi}{2} \end{aligned}$$

while

$$\begin{aligned} \iint_D \boldsymbol{u} \cdot \boldsymbol{n}\, dS &= \int_0^{2\pi} \int_0^{\sqrt{3}} \left(r^3 \cos\theta \sin^2\theta - 0\right) r\, dr\, d\theta \\ &= \left(\int_0^{2\pi} \cos\theta \sin^2\theta\, d\theta \right) \left(\int_0^{\sqrt{3}} r^4\, dr \right) = (0)(9\sqrt{3}/5) = 0 . \end{aligned}$$

Thus

$$\iint_H \boldsymbol{u} \cdot \boldsymbol{n}\, dS = \frac{9\pi}{2} .$$

6.4.2 Green's identities

The divergence theorem is not only used to compute various integrals, but it is also used to establish a number of other famous named results. Two of these have come to be known as Green's first and second identities. Green's first identity is a kind of higher-dimensional integration by parts and basically involves applying the divergence theorem to the quantity $u\boldsymbol{F}$ where u is a scalar function and $\boldsymbol{F}$ is a vector field. The classical version of this identity has $\boldsymbol{F} = \nabla v$ so that the divergence theorem is applied to $u\nabla v$.

Theorem 22 (Green's identities). *Suppose that u and v are both $C^2(\Omega)$ for some common domain Ω. Then Green's first identity is*

$$\int_{\Omega} (\nabla u \cdot \nabla v + u \bigtriangleup v)\, dV = \oint_{\partial\Omega} u\nabla v \cdot \boldsymbol{n}\, dS$$

where $\boldsymbol{n}$ is the outer *unit normal vector to $\partial\Omega$. Also Green's second identity is*

$$\int_{\Omega} (u \bigtriangleup v - v \bigtriangleup u)\, dV = \oint_{\partial\Omega} (u\nabla v - v\nabla u) \cdot \boldsymbol{n}\, dS\,.$$

Proof. The proof of Green's identities is discussed in Exercise 6.12. The key to proving Green's first identity is to apply the divergence theorem to the vector field $u\nabla v$. □

6.4.3 Stokes' theorem

Theorem 23 (Stokes). *Suppose that $S \subset \mathbb{R}^3$ is a bounded, connected, piecewise smooth, orientable surface, and let ∂S be the piecewise smooth boundary of S. Suppose that $\boldsymbol{u}$ is continuously differentiable up to the boundary: $\boldsymbol{u} \in C^1(\bar{S})$. Then*

$$\iint_{S} \operatorname{curl} \boldsymbol{u} \cdot \boldsymbol{n}\, dS = \oint_{\partial S} \boldsymbol{u} \cdot d\boldsymbol{x}$$

where ∂S is traversed in the positive direction relative to $\boldsymbol{n}$, a unit normal vector for S.

There is one new concept in Theorem 23: orientable surface.

Definition. A surface is *orientable* if and only if an ant standing on a point on the surface cannot walk on the surface to the same point but on the opposite side without crossing the boundary. For an orientable surface, its boundary ∂S is traversed in the positive direction (or orientation) if and only if the direction of motion along the boundary and the unit normal vector $\boldsymbol{n}$ satisfy the right-hand rule: if one curls the fingers of one's right hand in the direction of motion along the boundary, the right thumb points in the direction of $\boldsymbol{n}$ as is shown in Figure 6.6.

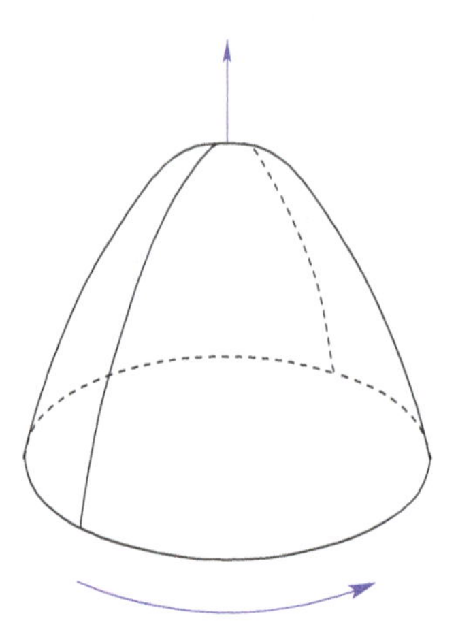

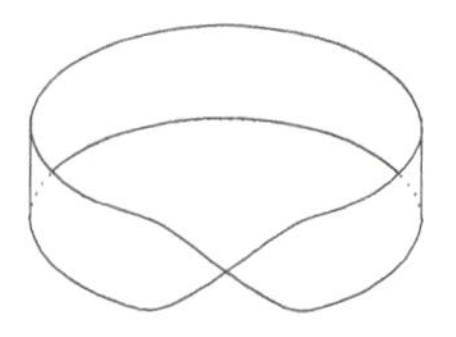

Figure 6.6: Postive Orientation (left) and Möbius Strip (right). If one curls the fingers of one's right hand in the direction indicated by the blue directed arc, one's right thumb points in the direction of $\boldsymbol{n}$. The Möbius strip is a band with one loop, but also with one twist.

At first glance, one might think that all smooth surfaces are orientable, but this is not true. The classic example of a nonorientable surface is a Möbius[5] strip. To make a Möbius strip, cut out a small strip of paper approximately 25 cm (10 inches) long and 5 cm (2 inches) wide. Next form a band by bringing the 5-cm ends of the band together, but instead of forming a simple band, insert one twist as shown in Figure 6.6. If one runs a finger along this strip, one can see that it has only one side.

Remarks.

1. If the surface S is embedded in $\mathbb{R}^2$, and $\boldsymbol{u} = \langle F, G \rangle$ (traditional notation), then $\boldsymbol{n} = \langle 0, 0, 1 \rangle$, and Stokes' theorem reduces to a form that is usually called Green's theorem:

$$\iint_S \left(\frac{\partial F}{\partial y} - \frac{\partial G}{\partial x} \right) dA = \oint_{\partial S} \langle F, G \rangle \cdot d\boldsymbol{x}$$

2. Physically and mathematically, what this theorem says that the total amount of circulation on the surface S is equal to the amount that the vector field flows along the boundary ∂S. Thus this theorem is again a statement of conservation of flow for the vector field, but now in the tangential or circulation sense. If the curl is positive, then on average the vector field flows along the boundary ∂S in the positive direction (according to the right-hand rule). If the curl is negative, then on average the vector field flows along the boundary ∂S in the opposite (negative) direction.
3. As was the case when surface integrals were defined, the version of Stokes' theorem given here does not directly cover closed surfaces (e. g., spheres) because our

5 Named for August Möbius (1790–1868), a German mathematician and astronomer.

definition of a smooth surface requires that the surface be explicitly representable as $z = f(x, y)$ where f is a differentiable function, and of course this is impossible for closed surfaces. But as before, for a sphere, the solution to this problem is simply to divide it into an upper hemisphere and a lower hemisphere. Also the line integrals around the boundaries (the equator in this case) are equal in magnitude and opposite in sign, hence they cancel. So for a sphere S,

$$\iint_S \operatorname{curl} \boldsymbol{u} \cdot \boldsymbol{n}\, dS = 0\,.$$

This result generalizes to closed surfaces (any surface with $\partial S = \emptyset$) that can be divided into smooth upper and lower portions, and indeed to any bounded, connected, piecewise smooth, orientable, closed surface.

4. As was the case for the divergence theorem above, this is not the most general statement of Stokes' theorem.

Corollary 4. *Suppose that $S \subset \mathbb{R}^3$ is a closed, bounded, connected, piecewise smooth, orientable surface. Then $\partial S = \emptyset$ and*

$$\oiint_S \operatorname{curl} \boldsymbol{u} \cdot \boldsymbol{n}\, dS = 0$$

where $\boldsymbol{n}$ is the outward unit normal vector for S and $\boldsymbol{u}$ is any continuously differentiable vector field.

As with the divergence theorem, Stokes' theorem can sometimes be used to turn a complicated integral into a much simpler integral.

Example 6.11. For the vector field $\boldsymbol{u}(x, y, z) = \langle -y, x, 1\rangle$ and for a conical silo surface S with a circular base whose radius is R (see Figure 6.7), please compute

$$\iint_S \operatorname{curl} \boldsymbol{u} \cdot \boldsymbol{n}\, dS$$

where $\boldsymbol{n}$ is the upward unit normal vector for S.

Answer. By direct computation, one finds that $\operatorname{curl} \boldsymbol{u} = \langle 0, 0, 2\rangle$, but it would be perhaps more difficult to carefully parameterize the surface of the silo. By Stokes' theorem, however, the requested surface integral can be found by computing the line integral

$$\oint_{\partial S} \boldsymbol{u} \cdot d\boldsymbol{x} = \int_0^{2\pi} \boldsymbol{u} \cdot \frac{d\boldsymbol{x}}{d\theta}\, d\theta = \int_0^{2\pi} \langle -R\sin\theta, R\cos\theta, 1\rangle \cdot \langle -R\sin\theta, R\cos\theta, 0\rangle\, d\theta$$

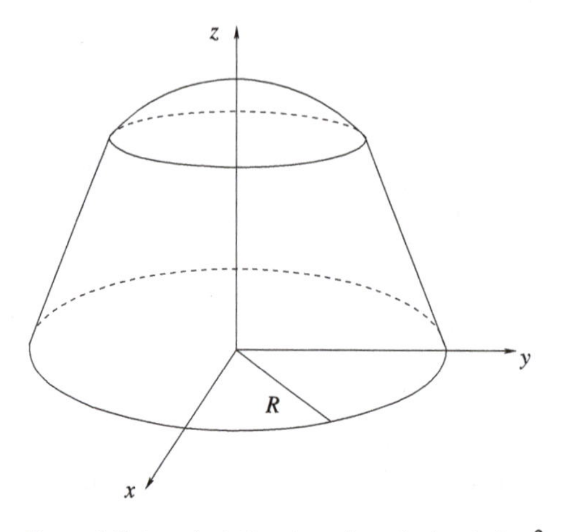

Figure 6.7: A conical silo whose base is the circle $x^2 + y^2 = R^2$.

$$= R^2 \int_0^{2\pi} (\sin^2\theta + \cos^2\theta)\, d\theta = 2\pi R^2 .$$

So Stokes' theorem implies that the exact shape of the silo is not important as long as the shape of the base is fixed. And indeed one could also use Stokes' theorem and Corollary 4 to just integrate the curl over the base of the silo:

$$\iint_S \text{curl } \boldsymbol{u} \cdot \boldsymbol{n}\, dS = -\iint_D \text{curl } \boldsymbol{u} \cdot \boldsymbol{n}\, dA = -\iint_D \langle 0,0,2\rangle \cdot \langle 0,0,-1\rangle dA = 2\text{Area}(D) = 2\pi R^2$$

where D is the disk of radius R at the base of the silo and for the integral over this base, $\boldsymbol{n} = \langle 0,0,-1\rangle$ is the downward unit normal vector.

Example 6.12. Suppose that $\boldsymbol{F} : \mathbb{R}^3 \to \mathbb{R}^3$ is a vector field, and suppose that there is a twice continuously differentiable function φ such that $\boldsymbol{F} = \nabla\varphi$. How does Stokes' theorem relate to this vector field?

Answer. Of course, as was discussed in Section 6.1.2, this vector field $\boldsymbol{F}$ is conservative, and so the line integral over any closed curve is zero. So for any surface S satisfying the conditions of Stokes' theorem, the integral around the closed boundary ∂S must be zero, and hence Stokes' theorem implies that

$$\iint_S \text{curl } \boldsymbol{F} \cdot \boldsymbol{n}\, dS = \iint_S \text{curl } \nabla\varphi \cdot \boldsymbol{n}\, dS = 0 .$$

But in fact, by direct computation, one finds that as an operator,

$$\text{curl } \nabla \equiv \boldsymbol{0}$$

regardless of which potential function this operator is applied to, as long as the potential function is twice continuously differentiable. So both the surface integral and the line integral from Stokes' theorem are zero.

6.5 All together now: a unified theorem

Up to this point, the theorems of Gauss, Green, and Stokes have been presented mostly as separate though clearly related results. In this concluding section, however, they are combined along with the fundamental theorem of calculus and given as a single result.

Theorem 24 (Green, Stokes, Cartan). *All of the above integral theorems of vector calculus along with the fundamental theorem of calculus can be expressed in a single form:*

$$\int_{\Omega} dF = \oint_{\partial\Omega} F$$

where, depending on setting, Ω is a bounded, open, connected domain in $\mathbb{R}^n$, or a bounded, connected portion of a smooth, orientable surface in $\mathbb{R}^3$; $\partial\Omega$ is the positively-oriented and piecewise smooth boundary of Ω; F is some continuously differentiable vector field or scalar function; and dF is the appropriate first derivative of F (divergence, curl, single-variable derivative).

Remarks.

1. The proof of this theorem is beyond the scope of this text and requires the concepts of differential forms and manifolds (see, e. g., Marsden and Tromba [2], Section 8.5). For our purposes, this result can be thought of as a notational summary of the theorems of Gauss, Stokes, and Green, and the fundamental theorem of calculus.
2. Notice that in this version, the theorem seems to say that one can commute the differential operator off the integrand and onto the domain of integration; this is essentially correct (again in the realm of differential forms) and justifies the use of the partial derivative symbol to denote the boundary of the open domain Ω.
3. Whose name should be on this theorem is difficult to say. Frequently, Stokes is credited with the theorem, but several of the vector-calculus forms of this result were due to Green, and the differential forms result is due to Cartan[6] (1945).

Example 6.13. What are Ω, $\partial\Omega$, F, and dF if Theorem 24 is to represent the fundamental theorem of calculus?

---

6 Élie Cartan (1869–1951) was a French mathematician who worked extensively on Lie groups and differential geometry mainly in the first half of the 20th century.

Answer. For the fundamental theorem of calculus, $\Omega = [a, b]$, an interval in the real line with $a < b$, the boundary $\partial\Omega = \{a, b\}$ is just the two end points, and $dF = f'(x)dx$ is the derivative of a continuously differentiable function $F = f$. In this case, the integral over the boundary is just evaluation at the endpoints, with the negative sign coming from outward unit normal “vector” at a being -1:

$$\int_a^b f'(x)\,dx = \int_{[a,b]} f'(x)\,dx = \int_\Omega dF = \oint_{\partial\Omega} F = f(b) - f(a)\,.$$

Exercises 6

6.1. Please compute each of the following line integrals:

(a) The line integral of $\boldsymbol{F}(x_1, x_2) = \langle x_1, x_2\rangle$ along the curve traced out by the vector function $\boldsymbol{x}(t) = \langle t, t\rangle$ from $(0, 0)$ to $(2, 2)$.

(b) The line integral of $\boldsymbol{F}(x, y, z) = \langle 2x, 3y^2, z\rangle$ along the curve traced out by the vector function $\boldsymbol{x}(t) = \langle t, 1/t, t^2\rangle$ from $(1, 1, 1)$ to $(2, 1/2, 4)$.

(c) The line integral of $\boldsymbol{F}(x, y, z) = \langle y, -z, x\rangle$ along the three-dimensional alpha curve traced out by the vector function $\boldsymbol{x}(t) = \langle t^2 - 1, t(t^2 - 1), t\rangle$ from $t = -2$ to $t = 2$.

(d) The integral of $\boldsymbol{F}(x, y) = \langle x, y\rangle$ over the unit circle from $(0, 0)$ to a point on the circle corresponding to the polar angle θ.

Answer. (a) 4; (b) 77/8; (c) 244/15; (d) 0

6.2. Use the definition of the line integral to prove Proposition 7 based on the similar result from single variable calculus:

$$\int_b^a f(x)\,dx = -\int_a^b f(x)\,dx$$

and

$$\int_a^c f(x)\,dx = \int_a^b f(x)\,dx + \int_b^c f(x)\,dx$$

provided that all of these integrals make sense.

6.3. Using the results of Examples 6.2 and 6.3, please compute the line integral

$$\int_S \boldsymbol{f}(\boldsymbol{x}) \cdot d\boldsymbol{x}$$

where $S = C - C_1$ is the closed semicircular loop starting at $(1, 0)$, moving along the upper unit circle to $(-1, 0)$, then moving along the x-axis back to $(1, 0)$.

6.4. For each of the following vector fields, please either find a potential function φ or determine that no such potential function exists. When a potential function exists, please compute the value I of any line integral from the origin (either $(0,0)$ or $(0,0,0)$) to either $(1,1)$ or $(1,1,1)$:

(a) $\boldsymbol{F}(x,y,z) = \langle\, yz + 2xe^z,\ xz,\ xy + x^2e^z \,\rangle$

(b) $\boldsymbol{F}(x,y,z) = \langle\, 4xy + e^z,\ 2x^2 + 2y,\ xe^z + 2yz \,\rangle$

(c) $\boldsymbol{f}(x_1,x_2) = \langle\, x_2e^{x_1x_2} + x_2,\ x_1e^{x_1x_2} + x_1 \,\rangle$

(d) $\boldsymbol{F}(x_1,x_2,x_3) = \langle\, \sin x_2 \cos x_3 + \cos x_1 \cos x_2,\ x_1 \cos x_2 \cos x_3 - \sin x_1 \sin x_2,\ x_1 \sin x_2 \sin x_3 \,\rangle$

Answer. (a) $\varphi(x,y,z) = xyz + x^2e^z$, $I = 1 + e$; (b) φ DNE; (c) $\varphi(x_1,x_2) = x_1x_2e^{x_1x_2} + x_1x_2$, $I = 1 + e$; (d) $\varphi(x_1,x_2,x_3) = x_1 \sin x_2 \cos x_3 + \sin x_1 \cos x_2$, $I = 2\sin 1 \cos 1$

6.5. Show that if $\boldsymbol{f} : \mathbb{R}^2 \to \mathbb{R}^2$ is a vector field, then $\boldsymbol{f}$ is conservative if and only if

$$\frac{\partial f_1}{\partial y} = \frac{\partial f_2}{\partial x}$$

where $\boldsymbol{f} = \langle f_1, f_2 \rangle$.

6.6. Its definition, Theorem 18 and Theorem 19 give three characterizations for a vector field being conservative; there is a fourth:

(a) Assuming that all derivatives exist, please show that $\operatorname{curl}(\nabla u) \equiv 0$ for any scalar function (scalar field) u.

(b) Please explain why a vector field $\boldsymbol{F}$ is conservative on its open domain if and only if $\operatorname{curl} \boldsymbol{F} \equiv 0$ on the domain of $\boldsymbol{F}$.

6.7. Suppose that a vector field $\boldsymbol{F}$ is conservative on some domain D, and pick a reference point $\boldsymbol{x}_o \in D$. Show that $\Phi(\boldsymbol{x}) := \int_{\boldsymbol{x}_o}^{\boldsymbol{x}} \boldsymbol{F}(\boldsymbol{\xi}) \cdot d\boldsymbol{\xi}$ is a potential function for $\boldsymbol{F}$, that is, $\boldsymbol{F}(\boldsymbol{x}) = \nabla\Phi(\boldsymbol{x})$.

6.8. Please evaluate the line integral

$$\int_E \boldsymbol{f}(x,y) \cdot \boldsymbol{n}\, ds$$

where E is the ellipse $4x^2 + y^2 = 4$, $\boldsymbol{n}$ is the outward unit normal vector, and $\boldsymbol{f}(x,y) = \langle x,y \rangle$. Hint: Notice that because this exercise is posed in the x,y-plane, it is possible to determine the unit normal vector $\boldsymbol{n}$ directly from the unit vector $\boldsymbol{T}$.

Answer. $2\int_0^{2\pi} \frac{dt}{\sqrt{1+3\cos^2 t}}$

6.9. For the vector field $\mathbf{F}(x,y) = \langle x,y \rangle$, please compute

$$\int_C \mathbf{F}(x,y) \cdot \boldsymbol{n}\, ds$$

where C is the curve $x = y^2$ in the x, y-plane from $(0, 0)$ to $(4, 2)$ and $\boldsymbol{n}$ is the downward and rightward unit normal vector.

Answer. $-8/3$

6.10. Please evaluate the surface integral

$$\iint_{\Pi} \boldsymbol{F} \cdot \boldsymbol{n}\, dS$$

where $\boldsymbol{F}(x, y, z) = \langle y^2, -z, -x^2 \rangle$, Π is the portion of the plane $x + 3y + z = 3$ lying in the first octant, and $\boldsymbol{n}$ is the upper unit normal vector for this plane.

Answer. $-13/2$

6.11. Please evaluate the surface integral

$$\iint_{S} \boldsymbol{F} \cdot \boldsymbol{n}\, dS$$

where $\boldsymbol{F}(x, y, z) = \langle z, y, x \rangle$, S is the sphere of radius 3 centered at the origin, and $\boldsymbol{n}$ is the outward unit normal vector for this sphere.

6.12. For the vector field $\boldsymbol{v}(x, y, z) = \langle xyz, xyz, xyz \rangle$ and the scalar function $u(x, y, z) = xyz$, please compute each of the following:

(a) $\operatorname{div}(\boldsymbol{v})$
(b) $\operatorname{curl}(\boldsymbol{v})$
(c) $\operatorname{div}(\operatorname{curl}(\boldsymbol{v}))$
(d) ∇u
(e) $\nabla(\operatorname{div}(\boldsymbol{v}))$
(f) $\operatorname{curl}(\nabla u)$

Answer. (a) $yz + xz + xy$; (b) $\langle\, x(z - y),\ y(x - z),\ z(y - x)\, \rangle$; (c) 0; (d) $\langle\, yz,\ xz,\ yz\, \rangle$; (e) $\langle\, z + y,\ x + z,\ x + y\, \rangle$; (f) 0

6.13. Please follow the general outline of the argument for Theorem 21 of a vector field to show that

$$\operatorname{curl}(\boldsymbol{u})(x, y) = \lim_{\epsilon \to 0} \frac{1}{|A|} \oint_{\partial A} \boldsymbol{u} \cdot d\boldsymbol{x}$$

that represents the circulation of a vector field.

6.14. Show that

$$\operatorname{div}(\boldsymbol{u})(x, y) = \lim_{\epsilon \to 0} \frac{1}{|D|} \oint_{\partial D} \boldsymbol{u} \cdot \boldsymbol{n}\, ds$$

where D is a disk centered at (x, y) of radius ϵ.

6.15. Consider the rectangular prism (Box) $B = [0,2]\times[0,3]\times[0,5]$ and the vector field $\boldsymbol{F}(x,y,z) = \langle x^2y, y^2z, z^2x\rangle$. Suppose that $\boldsymbol{n}$ is the outward unit normal vector. Compute

$$\oint_{\partial B} \boldsymbol{F}(x,y,z)\cdot \boldsymbol{n}\, dS$$

Answer. 465

6.16. Let P be the prism bounded in the first octant by the plane $\Pi : x + y + z = 1$ (so P is bounded by Π and the three coordinate planes). For $\boldsymbol{F}(x,y,z) = \langle x,y,z\rangle$, please directly compute

$$\iint_{\Pi} \boldsymbol{F}\cdot \boldsymbol{n}\, dS$$

where $\boldsymbol{n}$ is again the outward unit normal vector to P, then use the divergence theorem to compute this integral.

Answer. 1/2

6.17. Suppose that $u \in C^2(\Omega)$ for some open, bounded, connected domain $\Omega \in \mathbb{R}^n$, and suppose that $\triangle u = f$ for a given $f \in C(\Omega)$, that is f is continuous on Ω. Please compute

$$\oint_{\partial\Omega} \nabla u\cdot \boldsymbol{n}\, dS$$

Hint: $\triangle \equiv \operatorname{div}\nabla$. *Answer.* $\iint_{\Omega} f(x,y)\, dA$

6.18. Consider the solid hemisphere H bounded above by the surface $S\colon z = \sqrt{1-x^2-y^2}$ and below by the circular disk D centered at the origin with radius 1. For $\boldsymbol{F}(x,y,z) = \langle x,y,z\rangle$, please compute the surface integral

$$\iint_{S} \boldsymbol{F}(x,y,z)\cdot \boldsymbol{n}\, dS$$

where again $\boldsymbol{n}$ is the outward unit normal vector to S.

Answer. 2π

6.19. If $u(x,y) = \sin(xy)$ and $v(x,y) = x^2y$, please verify Green's first identity.

6.20.

(a) Please prove Green's first identity by applying the divergence theorem to $u\nabla v$ and proving a multivariable product rule:

$$\operatorname{div}(u\nabla v) = \nabla u\cdot\nabla v + u\triangle v$$

by directly using the scalar product rule.

(b) Prove Green's second identity by subtracting Green's first identity with u and v reversed from Green's first identity as stated.

6.21. For the vector field $F(x,y,z) = \langle z, x, y \rangle$, verify Stokes' theorem when the surface is the upper hemisphere $z = \sqrt{1 - x^2 - y^2}$ and its boundary is the unit circle in the x, y-plane.

6.22. For $u(x,y,z) = \langle 3z, 2x, y \rangle$, if S is the unit disk $x^2 + y^2 \leq 1$ lying in the plane $z = 3$, please compute

$$\oint_{\partial S} \boldsymbol{u} \cdot d\boldsymbol{x}$$

where the boundary is traversed in the positive direction.

Answer. 2π

6.23. Suppose that a vector field F is conservative everywhere in $\mathbb{R}^3$. Explain why for any bounded, connected, smooth, orientable surface $S \subset \mathbb{R}^3$, with a piecewise smooth boundary ∂S,

$$\iint_S \operatorname{curl} \boldsymbol{F} \cdot \boldsymbol{n} \, dS = 0 \, .$$

There are at least two possible explanations.

6.24. Please show by direct computation that when applied to vector fields on $\mathbb{R}^3$ that are twice continuously differentiable (meaning that all partial derivatives of first or second order exist and are continuous)

$$\operatorname{div} \operatorname{curl} \equiv 0 \, .$$

6.25. Determine what Ω, dF, $\partial\Omega$, and F from Theorem 24 are for both the divergence theorem and Stokes' theorem, as was done for the fundamental theorem of calculus in Example 6.13.

Bibliography

[1] W. A. J. Kosmala, *A Friendly Introduction to Analysis*, 2 ed., Pearson, Prentice Hall, Upper Saddle River, NJ, 2004.
[2] J. E. Marsden and A. Tromba, *Vector Calculus*, 6 ed., W.H. Freeman, New York, 2012.
[3] P. C. Matthews, *Vector Calculus*, Springer, London, 1998.
[4] M. Rosenlicht, *Introduction to Analysis*, Dover, New York, 1986.
[5] W. Rudin, *Principles of Mathematical Analysis*, 3 ed., McGraw-Hill, Inc., New York, 1976.

https://doi.org/10.1515/9783110660609-007

Index

Printed in the USA
CPSIA information can be obtained
at www.ICGtesting.com
JSHW061343130823
46466JS00001B/53

9 783110 660203